COMMON CORE
MATH 2
WORKBOOK

M000313988

prepaze

www.prepaze.com

Author: Ace Academic Publishing

Ace Academic Publishing is a leading supplemental educational provider committed to offering students an enjoyable and interactive learning experience. Through our comprehensive workbooks that are designed to include challenging, multi-step questions, we aim to provide students with state of the art educational materials that will help them improve their academic performance. Our carefully selected practice questions encourage logical thinking and creativity and combine the focus on the required common core standards along with the understanding of the practical applications of the mathematical concepts.

For inquiries, contact Ace Academic Publishing at the following address:

Ace Academic Publishing
3736 Fallon Road #403
Dublin CA 94568

www.aceacademicprep.com

Ace Academic Publishing
ACHIEVING EXCELLENCE TOGETHER

ISBN:978-1-949383-02-7

About the Book

The contents of this book includes multiple chapters and units covering all the required Common Core Standards for this grade level. Similar to a standardized exam, you can find questions of all types, including multiple choice, fill-in-the-blank, true or false, matching and free response questions. These carefully written questions aim to help students reason abstractly and quantitatively using various models, strategies, and problem-solving techniques. The detailed answer explanations in the back of the book help the students understand the topics and gain confidence in solving similar problems.

For the Parents

This workbook includes practice questions and tests that cover all the required Common Core Standards for the grade level. The book is comprised of multiple tests for each topic so that your child can have an abundant amount of tests on the same topic. The workbook is divided into chapters and units so that you can choose the topics that you want your child needs to focus on. The detailed answer explanations in the back will teach your child the right methods to solve the problems for all types of questions, including the free-response questions. After completing the tests on all the chapters, your child can take any Common Core standardized exam with confidence and can excel in it.

For additional online practice, sign up for a free account at www.aceacademicprep.com.

For the Teachers

All questions and tests included in this workbook are based on the Common Core State Standards and includes a clear label of each standard name. You can assign your students tests on a particular unit in each chapter, and can also assign a chapter review test. The book also includes two final exams which you can use towards the end of the school year to review all the topics that were covered. This workbook will help your students overcome any deficiencies in their understanding of critical concepts and will also help you identify the specific topics that your students may require additional practice. These grade-appropriate, yet challenging, questions will help your students learn to strategically use appropriate tools and excel in Common Core standardized exams.

For additional online practice, sign up for a free account at www.aceacademicprep.com.

www.prepaze.com

FOR ADDITIONAL PRACTICE AND HELP, VISIT OUR WEBSITE AT www.PREPAZE.com

YOU CAN FIND MORE WORKBOOKS FOR MATH AND ENGLISH FOR ALL GRADE LEVELS

   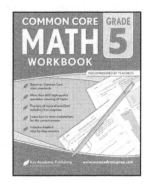

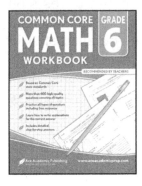  

TABLE OF CONTENTS

OPERATIONS & ALGEBRAIC THINKING

OPERATIONS & ALGEBRAIC THINKING

1. What number is 13 more than 6?

A. 18 **B.** 20 **C.** 15 **D.** 19

2.OA.B.2

2. John is picking 20 flowers. He has already picked 9 flowers. How many more flowers does he need to pick?

A. 11 **B.** 10 **C.** 13 **D.** 15

2.OA.B.2

3. Eighteen kids are playing on the playground. Later, 11 kids leave to go play in the sports field. How many kids are still on the playground?

A. 8 **B.** 9 **C.** 7 **D.** 11

2.OA.B.2

4. Jennie's mother packed her 15 grapes for lunch. She eats 6 of them. How many grapes does Jennie have left?

A. 9 **B.** 21 **C.** 6 **D.** 11

2.OA.B.2

5. Pedro has finished 11 math problems. His math homework has 19 questions on it. How many questions does Pedro have left to do?

A. 10 **B.** 8 **C.** 13 **D.** 2

2.OA.B.2

prepaze

OPERATIONS & ALGEBRAIC THINKING

ADD AND SUBTRACT WITHIN 20

6. Jessica's book has 14 chapters. She just finished reading Chapter 8. How many chapters does she have left to read?

    A. 12   **B.** 6   **C.** 7   **D.** 22

2.OA.B.2

7. John gave away 9 cans of food. He still has 7 cans of food on the shelf in his pantry. How many cans of food did he start with?

    A. 14   **B.** 2   **C.** 16   **D.** 9

2.OA.B.2

8. Jake has 9 books. His father buys him 12 more books. How many books does Jake have now?

    A. 27   **B.** 15
    C. 24   **D.** 21

2.OA.B.2

9. Julie is thinking of a number that is 13 less than the sum of 9 and 8. What number is she thinking of?

    A. 12   **B.** 10
    C. 6   **D.** 4

2.OA.B.2

10. There are eighteen students are in a classroom. When the bell rang, 5 students left to go to the music room.

How many students are in the classroom now?

    A. 9   **B.** 5   **C.** 13   **D.** 14

2.OA.B.2

OPERATIONS & ALGEBRAIC THINKING

ADD AND SUBTRACT WITHIN 20

11. Which equation could be used to find the answer to

13 + _____ = 20?

A. 13 − 20 = ?　　　　　　　　**B.** 20 − 13 = ?

C. 13 − 13 = ?　　　　　　　　**D.** 13 + 20 = ?

(2.OA.B.2)

12. Three cars are parked in the driveway. Each car has 4 wheels. How many wheels are in the driveway?

A. 4　　　　**B.** 8　　　　**C.** 3　　　　**D.** 12

(2.OA.B.2)

13. I am going to read 18 books this month. I read 4 books last week and 5 books this week. How many more books do I need to read?

A. 9　　　　**B.** 5　　　　**C.** 4　　　　**D.** 10

(2.OA.B.2)

14. Twenty students are on the playground. Two students go to the office and the rest go to lunch. How many students went to lunch?

A. 20　　　　**B.** 22　　　　**C.** 19　　　　**D.** 18

(2.OA.B.2)

15. Jessica has 17 ice cubes. She puts 13 ice cubes into her lemonade. How many ice cubes does Jessica have left?

A. 13　　　　**B.** 4　　　　**C.** 5　　　　**D.** 7

(2.OA.B.2)

prepaze

OPERATIONS & ALGEBRAIC THINKING

16. Ms. James takes out 17 balls for recess. Six of them are soccer balls and the rest are kickballs. How many kickballs did Ms. James bring to recess?

A. 6   **B.** 23   **C.** 11   **D.** 10

2.OA.B.2

17. My brother had 13 toy cars. He lost one toy car at the park but then he received 4 new toy cars for his birthday. How many toy cars does my brother have now?

A. 12   **B.** 16   **C.** 17   **D.** 18

2.OA.B.2

18. Lily had 8 flowers. She picked 12 more flowers from her garden. How many flowers does she have now?

A. 20   **B.** 21
C. 18   **D.** 4

2.OA.B.2

19. A bowl of 15 pieces of fruit is sitting on the table. The bowl has 8 apples and some oranges. How many oranges are in the bowl?

A. 15   **B.** 8
C. 9   **D.** 7

2.OA.B.2

20. Joseph is trying to earn 20 stickers on his sticker chart. Yesterday, he had 7 stickers. He earned 4 stickers today.

How many more stickers does Joseph need to fill up his chart?

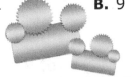

A. 11   **B.** 9   **C.** 7   **D.** 31

2.OA.B.2

UNIT 2: ADDITION AND SUBTRACTION WORD PROBLEMS

prepaze

OPERATIONS & ALGEBRAIC THINKING

1. Ms. Jacobs bought 2 boxes of 48 crayons for her class. How many crayons did she buy?

   A. 48   **B.** 88   **C.** 96   **D.** 86

   2.OA.A.1

2. April has a box with 95 folders in it. She needs to put 43 folders in Ms. Green's classroom and 37 folders in Mr. Anderson's classroom.

   How many folders will April have left?

   A. 80
   B. 95
   C. 15
   D. 14

   2.OA.A.1

3. On Monday, Allison checked out 13 storybooks for her siblings and 6 chapter books for herself from the library. On Wednesday, she returned 4 of the books.

   How many books does she still have checked out?

   A. 13   **B.** 23
   C. 9   **D.** 15

   2.OA.A.1

4. A bus picks up 16 children at the first stop and the same number at the second stop. At the third stop, the bus picks up 21 children.

   How many children are on the bus?

   A. 37   **B.** 11
   C. 36   **D.** 53

   2.OA.A.1

prepaze

OPERATIONS & ALGEBRAIC THINKING

5. Andrew has 84 stickers in his sticker book. Carlos has 62 stickers in his sticker book.

How many more stickers does Andrew have than Carlos has?

A. 84   **B.** 62   **C.** 22   **D.** 24

2.OA.A.1

6. Mr. Frank had 83 stickers at the beginning of the day. He gave out 24 stickers during reading and 17 during math.

Which equation represents this problem?

A. 83 + 24 = ?
B. 83 − 24 −17 = ?
C. 83 − 24 + 17 = ?
D. 83 + 24 + 17 = ?

2.OA.A.1

7. There are 3 second grade classes at my school. The first class has 26 students, the second class has 28 students, and the third class has 23 students.

How many second-grade students attend my school?

A. 77   **B.** 67   **C.** 54   **D.** 80

2.OA.A.1

8. Juan is collecting 90 shells for a project. He has collected 42 shells so far. How many more shells does he need to collect?

Juan needs _____ more shells.

2.OA.A.1

OPERATIONS & ALGEBRAIC THINKING

9. Right now, 53 children are swimming in the pool. fourteen children get out of the pool to eat lunch.

How many children are still in the pool?

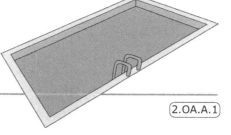

_____

2.OA.A.1

10. A car dealership has 82 cars to sell. On Monday they sold 14 cars. They also had 18 new cars delivered.

How many cars does the dealership have now?

The dealership has _____ cars.

Which operation sign goes on the blank line to make the equation represent the story?

$$82 - 14 \underline{\hspace{1cm}} 18 = ?$$

2.OA.A.1

11. April wants to bike 70 miles this week. She biked 28 miles on Monday and 14 miles on Tuesday.

How many more miles does she still need to bike this week to reach her goal?

April needs to bike _____ more miles.

2.OA.A.1

prepaze

OPERATIONS & ALGEBRAIC THINKING

ADDITION AND SUBTRACTION WORD PROBLEMS

12. An orchard has 48 orange trees, 16 lemon trees, and 28 lime trees. How many trees are in the orchard?

There are _____ trees in the orchard.

2.OA.A.1

13. A class of 30 students goes to the library. Each student can check out 3 books. How many books can they checkout in all?

The students can checkout _____ books in all.

2.OA.A.1

14. Sweets Bakery has sold 47 cupcakes today. They still have 38 cupcakes to sell. How many cupcakes did they start with?

Sweets Bakery started with _____ cup.

2.OA.A.1

15. Anthony is trying to collect 85 shells. He has already found 34 shells. How many more shells does he need to collect to reach his goal? Explain your reasoning.

_____

_____

_____

2.OA.A.1

OPERATIONS & ALGEBRAIC THINKING

16. There are 21 kindergarteners and 33 first graders waiting for lunch. There are 74 lunches in the cafeteria. How many lunches will be left after all the children eat? Explain your reasoning.

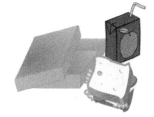

_____

_____

_____

2.OA.A.1

17. Tom had $90. He spends $62 at a grocery store. He wants to see a show that costs $20. Does he have enough money to buy a ticket for the show?

Write an equation to represent the problem. Explain how this equation could help you find the answer.

_____

_____

_____

2.OA.A.1

18. Edward brought 15 cookies and 25 muffins to a bake sale. Stephanie brought 24 cupcakes and 20 cookies. Who brought more treats to the bake sale? Explain your reasoning.

_____

_____

_____

2.OA.A.1

prepaze

OPERATIONS & ALGEBRAIC THINKING

19. Jessica's chapter book has 92 pages. She has 24 pages of her book left to read.

How many pages has she read so far? Explain what strategy you used to solve.

_____

_____

_____

(2.OA.A.1)

20. There are 45 red apples and 29 green apples. At lunch, 31 children each take one apple.

How many apples are left? Explain how you know.

_____

_____

_____

(2.OA.A.1)

UNIT 3:   FOUNDATIONS FOR MULTIPLICATION

OPERATIONS & ALGEBRAIC THINKING

1. Kayla is counting the number of guitars she sees in a music store by twos. How many sets of twos does Kayla see?

A. 7   **B.** 8   **C.** 9   **D.** 14

2.OA.C.3

2. Amanda put blueberries into 4 rows. Each row has 4 berries. How many blueberries does Amanda have?

A. 4   **B.** 8   **C.** 16   **D.** 20

2.OA.C.4

3. Alex and Alexa are sharing a bucket of toy cars, shown below. How many will each of them get?

A. 7   **B.** 9   **C.** 8   **D.** 16

2.OA.C.3

4. How many objects are in each column?

A. 2   **B.** 3   **C.** 4   **D.** 5

2.OA.C.4

prepaze

OPERATIONS & ALGEBRAIC THINKING

5. Frank is writing a doubles equation to go with this picture. Which equation should he write?

A. 5 + 5 = 10

B. 12 + 0 = 12

C. 5 + 6 = 12

D. 6 + 6 = 12

2.OA.C.3

6. Which repeated addition expression could be used to represent the bicycles in this picture?

A. 4 + 4 + 4

B. 5 + 5 + 5

C. 5 + 5

D. 3 + 3 + 3 + 3 + 3

2.OA.C.4

7. Sam drew the pictures below. Which equation represents the number of flowers Sam drew?

A. 5 + 5 = 10

B. 5 + 5 + 1 = 11

C. 5 + 5 = 11

D. 6 + 6 = 12

2.OA.C.3

OPERATIONS & ALGEBRAIC THINKING

8. Jake is writing an equation for this picture.

He should add the number _____ three times.

2.OA.C.4

9. Does Celeste have an even or odd number of stickers shown below?

Celeste has an _____ number of stickers.

2.OA.C.3

10. Sandra put her crayons into 2 columns and 3 rows. How many crayons does she have?

Sandra has _____ crayons.

2.OA.C.4

11. Ms. Thompson asked you to put the same number of the juice boxes, in the figure below, on the blue table and on the purple table. How many do you have left over?

I have _____ leftover.

2.OA.C.3

prepaze

FOUNDATIONS FOR MULTIPLICATION

OPERATIONS & ALGEBRAIC THINKING

FOUNDATIONS FOR MULTIPLICATION

12. I can find the total number of stars by adding the number 4 _____ times.

2.OA.C.4

13. Are there an even or odd number of bikes in this picture?

There are _____ number of bikes.

2.OA.C.3

14. Your teacher asks you to draw an array to match the equation

3 + 3 + 3 = 9. Describe what your drawing will look like.

_____

_____

_____

2.OA.C.4

prepaze

OPERATIONS & ALGEBRAIC THINKING

15. James is bagging the apples he picked today. He wants to give half of them to his mom and the other half to his uncle. How many apples should he put into each bag?

Jame should put _____ apples in each bag.

(2.OA.C.3)

16. How can Rebeca use repeated addition to help her count these guitars? Explain your reasoning.

_____

_____

_____

(2.OA.C.4)

17. Does Mark have an even or odd number of apples? How do you know?

_____

_____

_____

(2.OA.C.3)

OPERATIONS & ALGEBRAIC THINKING

FOUNDATIONS FOR MULTIPLICATION

18. Martin says that the repeated addition sentence 5 + 5 + 5 + 5 matches this picture.

Do you agree or disagree? Why or why not?

_____

_____

_____

(2.OA.C.4)

19. Michelle says she has an even number of colored pencils in this cup. Do you agree or disagree? Explain your reasoning.

_____

_____

_____

(2.OA.C.3)

OPERATIONS & ALGEBRAIC THINKING

20. Are there an even or odd number of birds in this picture? Explain your reasoning?

_____

_____

_____

2.OA.C.3

CHAPTER REVIEW

prepaze

OPERATIONS & ALGEBRAIC THINKING

1. Marcus and Jessica are playing basketball. Marcus made 11 baskets. Jessica made 14. How many more baskets did Jessica make?

A. 25   **B.** 4   **C.** 5   **D.** 3

2.OA.B.2

2. Ashley made 47 snowballs. Her sister made 32 snowballs. Ashley and her sister had a snowball fight and threw 41 snowballs at each other. How many snowballs do they have left?

A. 47   **B.** 79   **C.** 38   **D.** 41

2.OA.A.1

3. Which number goes in the blanks to make a doubles fact? _____ + _____ = 20

A. 7   **B.** 11
C. 9   **D.** 10

2.OA.C.3

4. There are 87 cars and 43 trucks in the parking lot. How many more cars are there than trucks?

A. 44   **B.** 43
C. 31   **D.** 45

2.OA.A.1

5. Daniel has two boxes of crayons. Each box has 8 crayons in it. Daniel gave 4 crayons to his sister. How many crayons does he have left?

A. 12   **B.** 16   **C.** 20   **D.** 19

2.OA.B.2

OPERATIONS & ALGEBRAIC THINKING

6. Which doubles expression represents this picture?

A. 5 + 5

B. 6 + 6

C. 7 + 7

D. 8 + 8

2.OA.C.3

7. Jacob has 9 cars and 8 trucks. How many vehicles does he have in all?

A. 17   **B.** 14   **C.** 16   **D.** 20

2.OA.B.2

8. Martin is counting by 2's to figure out how many cherries he has. Which doubles fact can he use to count his cherries?

A. 8 + 8

B. 6 + 6

C. 9 + 9

D. 7 + 7

2.OA.C.3

prepaze

OPERATIONS & ALGEBRAIC THINKING

CHAPTER REVIEW

9. Ashley was picking apples. She picked 19 apples to make applesauce and a pie. She used 7 apples to make applesauce and the rest of the apples to make the pie. How many apples did she use for the pie?

A. 26 **B.** 11 **C.** 12 **D.** 14

2.OA.B.2

10. Which repeated addition expression matches this picture?

A. 4 + 4 + 4 + 4
B. 5 + 5 + 5 + 5
C. 5 + 5 + 5
D. 3 + 3 + 3 + 3

2.OA.C.4

11. Noah has 16 stickers. Edwin has 9 stickers. How many stickers do they have together?

A. 27 **B.** 25 **C.** 7 **D.** 17

2.OA.B.2

12. Ryan is writing a repeated addition equation to go with this picture. How many times does he need to add the number 3?

A. 2 **B.** 3 **C.** 4 **D.** 5

2.OA.C.4

OPERATIONS & ALGEBRAIC THINKING

13. April is skip counting to figure out how many juice boxes are on the table. How should April count these juice boxes?

She should count _____, 8, 12.

2.OA.C.4

14. Maria and Steven are sharing a bin of 94 blocks. Maria took 42 blocks. Steven took 21 blocks. How many blocks are still in the bin? Explain your reasoning.

_____

_____

_____

2.OA.A.1

15. At school students are required to line their bikes up in rows. How many bikes are in each row?

There are _____ bikes in each row.

2.OA.C.4

prepaze

OPERATIONS & ALGEBRAIC THINKING

16. Justin has 37 math problems to do for homework tonight. He did 16 at school and then 17 before dinner.

How many problems does Justin have to do? Write an equation that can be used to answer this question. Use a question mark to represent the unknown number.

_____

_____

_____

2.OA.A.1

17. Are there an even or odd number of basketballs, shown in this picture, in the gym?

There are an _____ number of basketballs.

2.OA.C.3

18. Sy'iona and Edwin are sharing a sheet of stickers equally. How many will each of them get?

They should each get

_____ stickers.

2.OA.C.3

OPERATIONS & ALGEBRAIC THINKING

19. There are 92 second graders going on a field trip to the science center. They are riding on two busses. Fifty-one students ride on the first bus. How many second graders are on the second bus?

There are _____ second graders on the second bus.

(2.OA.A.1)

20. What repeated addition sentence could be written about these apples? How do you know?

_____

_____

_____

_____

(2.OA.C.4)

EXTRA PRACTICE

prepaze

OPERATIONS & ALGEBRAIC THINKING

1. In the first half of the basketball game, we scored 45 points. In the second half, we scored 41 points. Which equation shows the total number of points we scored?

   A. $45 - 41 = ?$ **B.** $? - 41 = 45$ **C.** $41 + ? = 45$ **D.** $45 + 41 = ?$

   2.OA.A.1

2. Jon is thinking of a number that is 8 more than 14 minus 3. What number is Jon thinking of?

   A. 18   **B.** 25   **C.** 23   **D.** 19

   2.OA.B.2

3. In a classroom, 13 children are on the rug and 7 children are sitting at tables. How many children are in the classroom?

   A. 20   **B.** 13   **C.** 21   **D.** 23

   2.OA.B.2

4. Daniel has 19 students in his class. He wants to make a card for each of them. He has 8 cards left to make. How many cards has he made so far?

   A. 11   **B.** 13   **C.** 9   **D.** 27

   2.OA.B.2

5. Raymondo is not sure how to figure out if a number is even or odd. Explain one strategy he could use.

   _____

   _____

   _____

   2.OA.C.3

OPERATIONS & ALGEBRAIC THINKING

6. Is there an odd or an even number of lemons in this group? Explain how you know.

_____

_____

_____

[2.OA.C.3]

7. Which doubles fact tells us how many flowers are in these two bouquets?

A. 4 + 6   **B.** 5 + 5

C. 3 + 7   **D.** 9 + 1

[2.OA.C.3]

8. Amber and Edward want to share these strawberries equally. How should they divide the strawberries?

A. They each get 6 and have 0 left over.

B. They each get 6 and have 1 left over.

C. They each get 7 and have 0 left over.

D. They each get 7 and have 1 left over.

[2.OA.C.3]

OPERATIONS & ALGEBRAIC THINKING

9. How many objects are in each row?

A. 2   **B.** 3   **C.** 4   **D.** 5

2.OA.C.4

10. There are 3 rows of trees in an orchard and each column has 4 trees in it. How many trees are in the orchard?

A. 12

B. 7

C. 10

D. 15

2.OA.C.4

11. Robert is drawing a picture for the expression 3+3+3+3. What should he draw?

A. 3 rows of 3 objects

B. 4 rows of 3 objects

C. 5 rows of 4 objects

D. 3 rows of 5 objects

2.OA.C.4

12. Ana went to the store with 73 cents. She bought a sticker for 50 cents. When she got home she found 13 cents on her dresser. How much money does she have now?

Ana has _____ cents.

2.OA.A.1

13. Jacob put his books into 5 columns and 5 rows. How many books does he have?

Jacob has _____ books.

2.OA.C.4

14. Tina is writing an equation to go with this picture. How many times should she add the number 4?

She should add the number 4 _____ times.

2.OA.C.4

OPERATIONS & ALGEBRAIC THINKING

15. Sam and Juan are dividing up some cherries, shown below. They both love cherries and want as many as possible, but their shares must be equal. How many cherries will they have left over?

They will have _____ left over.

(2.OA.C.3)

16. Jake wants to divide these cloud stickers into two equal groups. How many groups can he make?

Jake can group these stickers into two

equal groups of _____.

(2.OA.C.3)

17. Rachel hopes to raise $75 with her two-day fundraiser. On the first day, she raised $48. On the second day, she raised $31. Did she reach her goal? How do you know?

_____

_____

_____

(2.OA.A.1)

prepaze

OPERATIONS & ALGEBRAIC THINKING

18. Quan says that he can use the doubles fact 8 plus 8 to figure out how many cubes he has. Do you agree? Why?

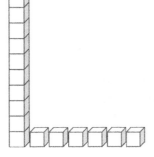

_____

_____

_____

(2.OA.C.3)

19. Scott says his chart has 4 columns and 2 rows. Do you agree or disagree? Why?

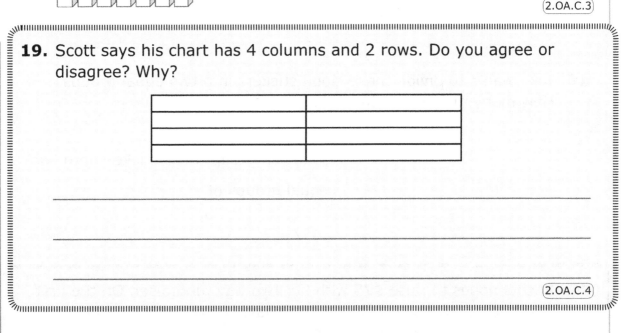

_____

_____

_____

(2.OA.C.4)

20. Jessica has 12 stickers. She wants to organize them into a rectangular array. Describe one rectangular array she could make.

_____

_____

_____

(2.OA.C.4)

NUMBER & OPERATIONS IN BASE TEN

prepaze

www.prepaze.com

NUMBER & OPERATIONS IN BASE TEN

1. In the number 704, what digit is in the tens place?

A. 7
B. 0
C. 4
D. 704

2.NBT.A.1

2. Which of these numbers is the same as 50 tens?

A. 5
B. 50
C. 500
D. 550

2.NBT.A.1

3. Alex has 8 hundreds blocks, 0 one blocks, and 7 tens blocks. What number does Alex have?

A. 870
B. 780
C. 807
D. 708

2.NBT.A.1

4. The expression 700 + 40 + 5 is the expanded form of what number?

A. 700
B. 745
C. 754
D. 7450

2.NBT.A.3

5. Jack counted 639 cars on his road trip. If he counts one more car, how many cars will he have counted?

A. 638   **B.** 640   **C.** 641   **D.** 649

2.NBT.A.2

prepaze

NUMBER & OPERATIONS IN BASE TEN

UNDERSTANDING PLACE VALUE

6. Which equation shows the expanded form of four hundred ninety-three?

A. 400 + 93 **B.** 490 + 3

C. 400 + 90 + 3 **D.** 400 + 9 + 3

2.NBT.A.3

7. Raquel has 890 base ten cubes. She gets one more ten stick. How many base ten cubes does she have?

A. 880 **B.** 895 **C.** 900 **D.** 910

2.NBT.A.2

8. What number does Samantha have in these blocks?

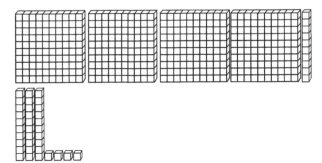 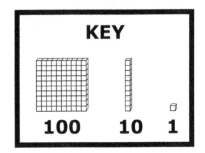

A. Four hundred forty **B.** Four hundred fourteen

C. Four hundred four **D.** Four hundred forty-four

2.NBT.A.3

9. Which symbol goes in the blank to make the statement true?

Nine hundred seventeen _____ 70 + 900

A. < **B.** = **C.** ! **D.** >

2.NBT.A.4

NUMBER & OPERATIONS IN BASE TEN

10. James is writing a number that is equal to the blocks shown in this picture. Which number should he write?

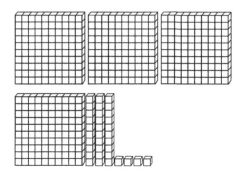

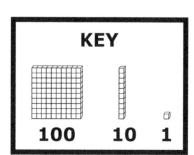

A. 444   **B.** 434   **C.** 334   **D.** 433

(2.NBT.A.4)

11. Which symbol goes in the blank to make the statement true?

Eight hundred forty-five _____ 844

A. <   **B.** !   **C.** >   **D.** =

(2.NBT.A.4)

12. Jessica says that she has ten tens of cubes. How many cubes does Jessica have? Explain your reasoning.

_____

_____

_____

_____

(2.NBT.A.1)

NUMBER & OPERATIONS IN BASE TEN

UNDERSTANDING PLACE VALUE

13. Mark is counting by 5's to 1000. He just said 860. What number should he say next? How do you know?

_____

_____

_____

2.NBT.A.2

14. Write the number eight hundred eighty in number form.

_____

2.NBT.A.3

15. Edwin has 7 tens 8 hundreds and 9 ones. Scott has 9 + 80 + 700 cubes. Who has a greater number?

_____

2.NBT.A.4

16. Sean is counting by 100's. What should he say after 893? Explain your reasoning.

_____

_____

_____

2.NBT.A.2

NUMBER & OPERATIONS IN BASE TEN

17. Is the following inequality true or false?

Eight hundred fifty-two< 3 + 40 + 800

_____

2.NBT.A.4

18. James is writing these base ten blocks using expanded form. He writes "200 + 40 + 2" do you agree or disagree? Why?

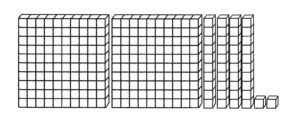
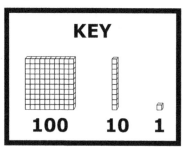

KEY

100   10   1

_____

_____

_____

_____

2.NBT.A.3

19. What number is 4 tens 8 ones and 3 hundreds?

_____

2.NBT.A.1

prepaze

NUMBER & OPERATIONS IN BASE TEN

UNDERSTANDING PLACE VALUE

20. How many base ten blocks does Samuel have?

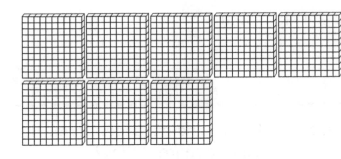

KEY		
100	10	1

Samuel has _____ blocks.

2.NBT.A.2

NUMBER & OPERATIONS IN BASE TEN

1. Mrs. Wolf had 37 pencils. She gave some pencils away. Now she has 23 pencils. Which number sentence can be used to find how many pencils Mrs. Wolf gave away?

   A. 37 − _____ = 23   **B.** 23 − 37= _____

   C. 23 + 37=_____   **D.** 23 − _____= 37

   2.NBT.B.5

PROPERTIES OF OPERATIONS

2. This graph shows the distance Mr. Ono travels each day.

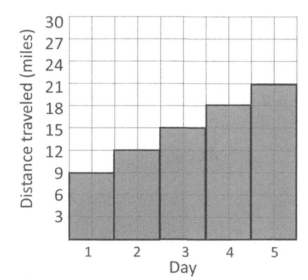

If the pattern shown on this graph continues, how many total miles will he have traveled by the end of Day 6?

   A. 21
   B. 24
   C. 99
   D. 75

   2.NBT.B.5

3. Which expression is modeled on this number line?

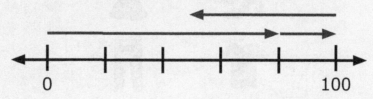

   A. 40+10−25   **B.** 80+20−50
   C. 80+20−40   **D.** 100+20−65

   2.NBT.B.5

prepaze

NUMBER & OPERATIONS IN BASE TEN

PROPERTIES OF OPERATIONS

4. This table shows the amount of money Donald saves each month.

Donald's Bank Account	
Month	**Money Saved (Dollars)**
February	9
March	13
April	17
May	21

If amount Donald saves continues with the same pattern each month, how much total money will he have saved by June?

A. $60

B. $25

C. $64

D. $85

2.NBT.B.5

---

5. There are 46 books on a shelf. On Monday, the students in Mrs. Jones' class borrow 22 books. On Friday, the students return 11 books. How many books are on the shelf on Friday?

There are _____ books on the shelf.

2.NBT.B.5

---

6. Ms. Wilson buys these items at the store. If she gives the cashier $100, how much money will she receive back?

Ms. Wilson will receive _____ dollars.

2.NBT.B.5

NUMBER & OPERATIONS IN BASE TEN

7. Maya is playing this board game. She currently has 60 points. She rolls the dice and moves her game piece 3 spaces ahead. What is Maya's score now?

	Add 20	Subtract 15	Add 35	Add 5	FINISH

Maya has _____ points.

2.NBT.B.5

8. Angel reads a book for 25 minutes, watches television for 36 minutes, then reads a different book for 18 minutes. How many more minutes does Angel spend reading than watching television?

_____ minutes

2.NBT.B.5

9. This thermometer shows the temperature in Mark's classroom. If the temperature in the classroom increases by 8 degrees, then decreases by 4 degrees, what is the new temperature?

The temperature in the classroom

is _____ degrees Fahrenheit

2.NBT.B.5

prepaze

NUMBER & OPERATIONS IN BASE TEN

PROPERTIES OF OPERATIONS

10. A goat travels from the bottom to the top of this mountain and back each day. How many miles does the goat travel after 9 days? Show your work.

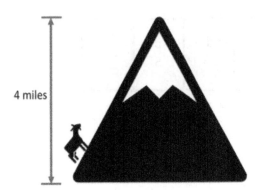

4 miles

2.NBT.B.5

11. These tables show the number of points scored by members of a basketball team in two different games.

Game 1	
Player	**Points**
Tom	12
Vaughn	18
Joe	22
Kola	10

Game 2	
Player	**Points**
Tom	15
Vaughn	17
Joe	24
Kola	19

How many more points did the team members score in Game 2 than Game 1?

A. 62   **B.** 137   **C.** 75   **D.** 13

2.NBT.B.6

NUMBER & OPERATIONS IN BASE TEN

12. Eric reads a book with 4 chapters. This table shows how many pages are in each chapter.

Chapter	Number of Pages
1	17
2	16
3	21
4	14

How many pages does he read altogether?

A. 86 pages   **B.** 68 pages
C. 58 pages   **D.** 618 pages

2.NBT.B.6

13. Josie exercises 45 minutes on Monday. She exercises 20 minutes more on Wednesday than she does on Friday. She exercises 15 minutes on Friday. How many minutes does Josie exercise on Monday, Wednesday, and Friday?

A. 70 minutes   **B.** 90 minutes
C. 710 minutes   **D.** 95 minutes

2.NBT.B.6

14. There are 15 dogs, 12 kittens, and 15 lizards in a pet shop. Which number sentence can you use to find the total number of animals in the pet shop?

A. $15 + 12 - 15$ **B.** $15 + 12 + 15$
C. $12 - 15 - 15$ **D.** $15 - 12 + 15$

2.NBT.B.6

NUMBER & OPERATIONS IN BASE TEN

15. This table shows the number of points scored by members of a basketball team in their last game.

Player	Points
Corey	20
Lexi	17
Hank	21
Mike	19

Which two players scored a total of 40 points?

_____

2.NBT.B.6

16. Nori buys 3 items from this store and spends $72.

$18

$35

$10

$8

$17

$29

$54

Which items does she buy?

_____

2.NBT.B.6

17. John is 15 years old. John's mother is 28 years older than him. John's father is 10 years older than his mother.

How old is John's father?

_____

2.NBT.B.6

NUMBER & OPERATIONS IN BASE TEN

18. This graph shows the number of students in 3 classrooms.

Mr. Lyons

Mrs. Simon

Mr. Jackson

= 10 students

How many students are in Mrs. Simon's classroom? _____

How many students are there altogether? _____

2.NBT.B.6

19. Tory, Sam, and Ronaldo are in a bike race.

- Tory finishes the race in 30 minutes.

- Sam finishes the race 7 minutes before Tory.

- Ronaldo finishes the race 12 minutes after Sam.

How much total time do all 3 of them ride to finish the race?

_____ minutes

2.NBT.B.6

prepaze

PROPERTIES OF OPERATIONS

NUMBER & OPERATIONS IN BASE TEN

PROPERTIES OF OPERATIONS

20. This number line models the distance a bird flies as it travels across the country.

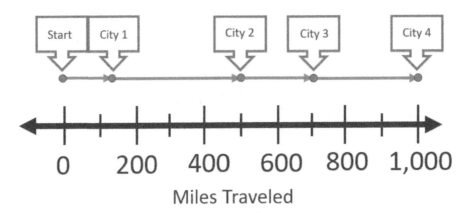

Miles Traveled

Write a number sentence to represent the distance the bird travels between the second and fourth cities.

_____

(2.NBT.B.6)

UNIT 3: PLACE VALUE TO ADD AND SUBTRACT

prepaze

NUMBER & OPERATIONS IN BASE TEN

1. Start at 55 on this number line. Make 3 hops to the right of 25 each.

55

What number are you on now?

A. 75   **B.** 100   **C.** 105   **D.** 130

2.NBT.C.7

2. Start at 110 on this number line. Make 3 hops to the right of 150 each.

110

What number are you on now?

A. 560   **B.** 160   **C.** 450   **D.** 260

2.NBT.C.7

3. The rule for this table is "add 99". Fill in the blanks.

In	Out
129	_____
358	_____
599	_____
901	_____

2.NBT.C.7

prepaze

NUMBER & OPERATIONS IN BASE TEN

ADDITION AND SUBTRACTION WORD PROBLEMS

4. A company creates this picture of 3 buildings they plan to construct.

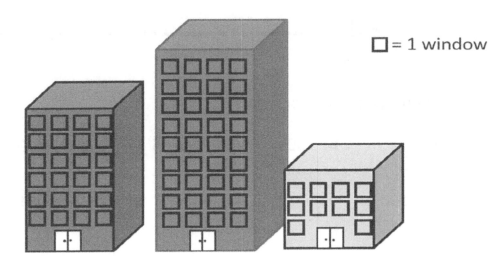

☐ = 1 window

If each building has the same number of windows on 3 sides, how many total windows are needed?

_____ windows are needed.

(2.NBT.C.7)

5. Model this subtraction equation on the number line:

$$621 - 388 = \underline{\hspace{3cm}}$$

0

(2.NBT.C.7)

NUMBER & OPERATIONS IN BASE TEN

6. Tony is planting flowers in this pot.

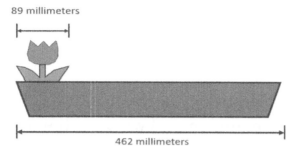

89 millimeters

462 millimeters

Write an addition number sentence to represent how much space Tony has left to plant more flowers in this pot.

_____

Write a subtraction number sentence to show how many more flowers of the same size he can add to this pot.

_____

2.NBT.C.7

7. Use the blocks to draw a model of this equation in the space below.

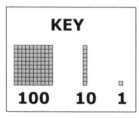

KEY

100   10   1

$135 + 462 =$ _____

2.NBT.C.7

NUMBER & OPERATIONS IN BASE TEN

ADDITION AND SUBTRACTION WORD PROBLEMS

8. ***Use mental math.*** In 10 years, Milsap Elementary School will be 150 years old. How old is the school now?

    A. 160   **B.** 140   **C.** 149   **D.** 151

    2.NBT.C.8

9. ***Use mental math.*** Mrs. Johnson owns a table that will be 275 years old 100 years from now. How old is the table now?

    A. 375 years   **B.** 274 years   **C.** 175 years   **D.** 75 years

    2.NBT.C.8

10. ***Use mental math.*** Lori has $325. She spends $10 at the movies, and $10 at the store. How much money does Lori have left?

    Lori has _____ dollars left.

    2.NBT.C.8

11. ***Use mental math.*** There is a total of 225 students in the second and third grades at Wagstaff Elementary. One hundred students are in the second grade. How many students are in the third grade?

    There are _____ students in the third grade.

    2.NBT.C.8

12. ***Use mental math.*** Randy and Paul eat snacks at the movies. These are the snacks Paul eats. Randy's snacks contain 100 more calories. How many calories are in Randy's snack?

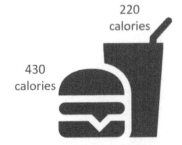

220 calories

430 calories

There are _____
calories in Randy's snack.

2.NBT.C.8

NUMBER & OPERATIONS IN BASE TEN

13. *Use mental math.* Tao and Mari are eating snacks. Tao eats 3 cookies, which have 100 calories each. Mari eats 1 more cookie than Tao. How many more calories does Mari eat?

Mari ate _____ calories more than Tao.

(2.NBT.C.8)

14. *Use mental math.* Ten ounces of water are added to a container containing 190 ounces of water. How much water is in the container now?

The container has _____ ounces of water in it.

(2.NBT.C.8)

15. Ben asks the students in his school to name their favorite food.

- 45 people like pizza
- 100 people like chicken nuggets
- 86 people like tacos

Which equation can Ben use to find how many more people like chicken nuggets and pizza than tacos?

A. $80 + 40 + (6 + 5) - 100 =$ _____

B. $100 + 45 =$ _____

C. $(100 + 40 + 5) - 80 - 6 =$ _____

D. $100 - (80 + 40) + (5 + 6) =$ _____

(2.NBT.C.9)

prepaze

NUMBER & OPERATIONS IN BASE TEN

ADDITION AND SUBTRACTION WORD PROBLEMS

16. Mora creates this number pattern by adding.

5, 8, 11, 14, 17, _____

What are the next two numbers in this pattern?

2.NBT.C.9

17. Brett creates this number pattern by adding.

10, 35, 60, 85, _____

What are the next two numbers in this pattern?

2.NBT.C.9

18. Joe and Tao are swimming. Joe swims 10 minutes longer than Tao. Tao swims for 32 minutes. Explain how you would determine how many minutes in total Joe and Tao spend swimming.

_____

_____

_____

2.NBT.C.9

19. A dog is at the park for 75 minutes. He runs around in the park for 16 minutes, takes a 30 minute nap, and the rest of the time digs holes. Does the dog spend more time taking a nap or digging holes? Why?

_____

_____

_____

2.NBT.C.9

NUMBER & OPERATIONS IN BASE TEN

20. Rex creates this number pattern by adding.

125, 200, 275, 350, 425

What number will be next in this pattern? Why?

_____

_____

_____

2.NBT.C.9

CHAPTER REVIEW

prepaze

NAME: _____ DATE: _____

NUMBER & OPERATIONS IN BASE TEN

CHAPTER REVIEW

1. The high score of Thalia's game is a 3-digit number. There is a 3 in the tens place, an 8 in the ones place, and a 1 in the hundreds place. What is the final score? How do you know?

_____

_____

_____

(2.NBT.A.1)

2. The number 700 is:

A. seven tens and zero ones   **B.** seven hundreds

C. seventy tens and ten ones   **D.** seven hundreds and one ten

(2.NBT.A.1)

3. Tom is writing the number three hundred forty. What digit should he write in the ones place?

Tom should write _____ in the ones place.

(2.NBT.A.1)

4. An orchard sold 450 bags of apples yesterday and 370 bags of apples today. How many bags of applies did the orchard sell all together?

A. 800   **B.** 802   **C.** 820   **D.** 810

(2.NBT.A.2)

5. Eight hundred forty tickets were sold for a concert online. Five more tickets were sold at the door. How many tickets were sold?

A. 850   **B.** 845   **C.** 855   **D.** 835

(2.NBT.A.2)

NUMBER & OPERATIONS IN BASE TEN

6. Fill in the blank in the counting sequence.

_____ , 715, 720, 725

(2.NBT.A.2)

7. Which number is nine hundred eighty-nine?

A. 989 **B.** 898 **C.** 988 **D.** 998

(2.NBT.A.3)

8. Which expression shows the number eight hundred forty-four in expanded form?

A. 800 + 4 + 4 **B.** 400 + 80 + 8
C. 800 + 44 **D.** 800 + 40 + 4

(2.NBT.A.3)

9. Max says that, "784 > 857 is correct because 784 is less than 857". Do you agree or disagree? Explain your reasoning.

_____

_____

_____

(2.NBT.A.4)

10. Which symbol makes this statement true?

600 + 40 + 5 _____ 654

A. < **B.** ! **C.** > **D.** =

(2.NBT.A.4)

prepaze

NUMBER & OPERATIONS IN BASE TEN

11. The table shows the height of 4 students in Mr. Chavez's class.

Student	Height (inches)
Javier	50
Sebastian	46
Mia	48
Jocelynn	49

What is the difference between the heights of the tallest student and shortest student?

_____ inches

2.NBT.B.5

12. This balance measures the mass of 2 identical toy cars in grams.

This balance measures the mass of a toy truck in grams.

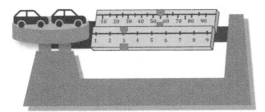
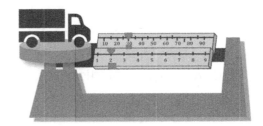

What is the mass of a toy car and a toy truck? Explain your reasoning.

_____

_____

_____

_____

2.NBT.B.5

NUMBER & OPERATIONS IN BASE TEN

13. Luke buys 4 items from this store and spends less than $65.

Which items does Luke buy?

_____

_____

(2.NBT.B.6)

14. Mondi is adding stamps to 2 pages in her scrapbook. Then she wants to add the same number of stamps on 4 more pages. How many stamps does she add altogether?

Mondi adds _____ stamps.

(2.NBT.B.6)

15. What is the rule for this table?

In	Rule	Out
257	?	445
438	?	626
576	?	764
794	?	982

A. Subtract 212   **B.** Add 212

C. Add 188   **D.** Subtract 188

(2.NBT.C.7)

prepaze

NUMBER & OPERATIONS IN BASE TEN

CHAPTER REVIEW

16. Start at 150 on this number line. Subtract 90 and then add 30 more to the result.

150

What number are you on now? _____

(2.NBT.C.7)

17. *Use mental math.* Nico has 56 books. His mother and grandmother each give him 10 more books. How many books does Nico have?

Nico has _____ books.

(2.NBT.C.8)

18. *Use mental math.* Roni collects 17 rocks. Her brother collects 100 rocks more than she does. Roni's sister collects 10 less than her brother. How many rocks does Roni's sister collect?

Roni's sister collected _____ rocks.

(2.NBT.C.8)

19. Lau is counting the number of fish in the pet store. This table shows her results.

Fish	Goldfish	Clownfish	Beta	Minnows
Total	87	94	66	102

...question 19. continued next page

NUMBER & OPERATIONS IN BASE TEN

She uses this strategy to find the total number of fish.

> *I rounded each number down to the next ten and decided to add 80 + 90 + 60 + 100 first. 80 + 90 + 60 + 100 = 230*
>
> *Then I add the digit in the ones place: 7 + 4 + 6 + 2.*
>
> *7 + 4 + 6 + 2 = 19*
>
> *Last, I add 230 + 19. The total number of fish is 249.*

Do you agree with Lau? Explain your reasoning.

_____

_____

_____

2.NBT.C.9

20. Willie has three boxes.

- ▫ The largest box weighs 52 pounds.
- ▫ The middle box weighs 8 pounds more than the smallest box.
- ▫ The weight of all three boxes is 94 pounds.

Explain how you would find the weight of the smallest and middle boxes.

_____

_____

_____

2.NBT.C.9

EXTRA PRACTICE

prepaze

NUMBER & OPERATIONS IN BASE TEN

1. Edwin is trying to draw the number 371 with base ten blocks but he is not sure how to do it. What would you tell him?

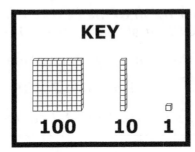

_____

_____

_____

(2.NBT.A.1)

2. James needs 294 cubes. How many more cubes does he need to add to the model below? Explain why.

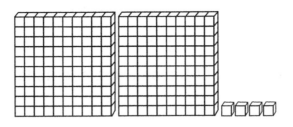   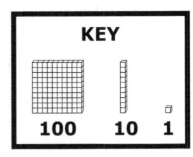

_____

_____

_____

(2.NBT.A.1)

NUMBER & OPERATIONS IN BASE TEN

3. Matt is thinking of a number that has 9 tens, 0 hundreds, and 4 ones. What number is Matt thinking of?

A. 904 **B.** 409 **C.** 94 **D.** 940

2.NBT.A.1

4. When counting by 5's what number comes immediately before 500?

_____

2.NBT.A.2

5. Scott bought 9 packs of paper. Each pack of paper has one hundred sheets in it. How many sheets of paper did he buy?

Scott bought _____ sheets of paper

2.NBT.A.2

6. Sara is counting by hundreds. What number will she say three numbers after 400?

A. 300 **B.** 500 **C.** 600 **D.** 700

2.NBT.A.2

7. Your partner is writing the number 625 in expanded form. He wrote "600 + 25". Do you agree or disagree? Why?

_____

_____

_____

2.NBT.A.3

prepaze

NUMBER & OPERATIONS IN BASE TEN

8. Jake is thinking of a number that has the expanded form 800 + 0 + 6. What number is he thinking of?

A. 860   **B.** 806   **C.** 80006   **D.** 8006

(2.NBT.A.3)

9. Which number is less than the cubes shown?

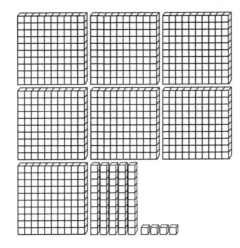

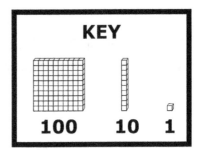

A. Eight hundred twenty-five   **B.** 6 + 40 + 700
C. 792   **D.** Seven hundred sixty

(2.NBT.A.4)

10. Fill in the missing symbol (<, > or =) to compare the two numbers.

600 + 40 + 0 _____ six hundred forty

(2.NBT.A.4)

11. Cleo's dog weighs 10 more pounds than Larry's dog. Larry's dog weighs 15 pounds less than Tommy's dog. Tommy's dog weighs 75 pounds. How much does Cleo's dog weigh?

A. 70 pounds   **B.** 65 pounds   **C.** 60 pounds   **D.** 100 pounds

(2.NBT.B.5)

NUMBER & OPERATIONS IN BASE TEN

12. Which strategy describes the first step for solving this number sentence?

$$37-24+14+6 = \underline{\hspace{3cm}}$$

A. Subtract 24 from 37

B. Add 14 and 24

C. Subtract 6 from 37

D. Add 14 and 6

(2.NBT.B.5)

13. Mark has 15 candies. Sara has 22 candies. Lance has 14 candies. Tim has 20 more candies than Mark. How many candies do they have altogether?

A. 86 candies　　**B.** 71 candies　　**C.** 61 candies　　**D.** 31 candies

(2.NBT.B.6)

14. Barry, Aja, Chuck, and Eric each have one of these gift boxes. The number on each box tells how many toys are inside.

▫ Barry's box has less than 20, but more than 16 toys.

▫ Aja's box has more than 12, but less than 15 toys.

▫ Chuck's box has more than 25 toys.

▫ Eric's box has less than 14 toys.

Which box belongs to each person?

Barry _____　　Aja _____　　Chuck _____　　Eric _____

How many toys do they have altogether? _____

(2.NBT.B.6)

prepaze

NUMBER & OPERATIONS IN BASE TEN

15. Start at 920 on this number line. Make three backward (left) hops of 240. Where are you on the number line?

0   920

A. 720   **B.** 680   **C.** 1,640   **D.** 200

2.NBT.C.7

16. Macy is trying to find a new pot for her flowers. The new pot should be 125 millimeters taller and 108 millimeters wider than this pot.

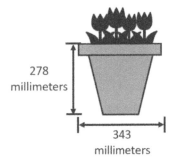

278 millimeters

343 millimeters

What are the height and width of the new pot?

Height: _____ millimeters

Width: _____ millimeters

2.NBT.C.7

17. *Use mental math.* There are 100 men, 100 women and some children at the park. If there are 375 people at the park, how many of them are children?

There are _____ children.

2.NBT.C.8

18. *Use mental math.* Luke has 5 one hundred-dollar bills. He gets some change and gives 10 dollars to his friend and spends 10 dollars at the store.

How much money does Luke have left?

Luke has _____ dollars left.

2.NBT.C.8

NUMBER & OPERATIONS IN BASE TEN

19. Rina uses a number line to solve this equation.

$$719 - 105 + 200$$

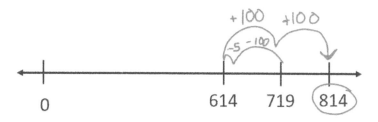

Do you agree with her strategy? Explain your reasoning.

_____

_____

_____

2.NBT.C.9

20. How are the rounding strategies for solving these equations similar?

79 + 100 = _____	400 − 128 = _____

Explain your reasoning.

_____

_____

_____

2.NBT.C.9

prepaze

MEASUREMENT & DATA

prepaze

MEASUREMENT & DATA

1. Tam is measuring this block with a centimeter ruler.

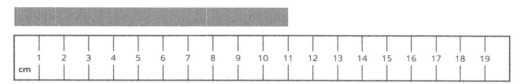

What is the length of the block?

A. 19 cm   **B.** 11 cm   **C.** 12 cm   **D.** 10 cm

(2.MD.A.6)

2. Which statement describes the length of the Usain's toy trumpet?

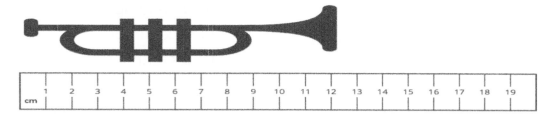

A. The length of the trumpet is (13 − 3) cm.

B. The length of the trumpet is (10 + 3) cm.

C. The length of the trumpet is (19 − 7) cm.

D. The length of the trumpet is (9 + 2) cm.

(2.MD.A.6)

3. How much longer is the black block than the white block?

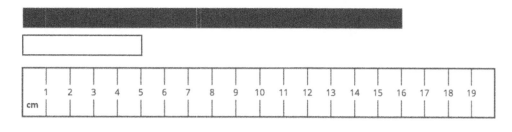

_____ cm

(2.MD.A.6)

NAME: .... DATE: ....

MEASUREMENT & DATA

MEASURING LENGTH

4. Which expression represents the length shown on this number line?

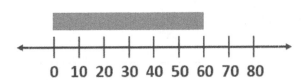

A. 80 – 60 units

B. 15 + 35 units

C. 80 – 20 units

D. 20 + 30 units

2.MD.A.6

5. What is the measure of the length of the sword to the nearest centimeter?

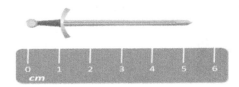

A. 3 centimeters

B. 4 centimeters

C. 6 centimeters

D. 5 centimeters

2.MD.A.1

6. What is the measure of the length of the crown to the nearest centimeter?

A. 6 centimeters   B. 4 centimeters

C. 2 centimeters   D. 3 centimeters

2.MD.A.1

7. What is the measure of the length of the dog, including the tail, to the nearest centimeter?

A. 6 centimeters

B. 5 centimeters

C. 4 centimeters

D. 6 centimeters

2.MD.A.1

prepaze

MEASUREMENT & DATA

8. How many dice or paper clips are needed to measure the line below?

_____

A. 1 dice or 2 paper clips

B. 2 dice or 1 paper clip

C. 3 dice or 1 paper clip

D. 0 dice or 2 paper clips

2.MD.A.2

9. How many dice or buttons are needed to measure the line below?

_____

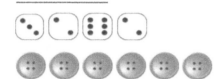

A. 1 dice or 1 button

B. 2 dice or 2 buttons

C. 3 dice or 1 button

D. 3 dice or 3 buttons

2.MD.A.2

10. How many buttons or paper clips are needed to measure the line below?

_____

A. 1 button or 5 paperclips   **B.** 2 buttons or 2 paper clips

C. 6 buttons or 3 paper clips   **D.** 5 buttons or 4 paper clips

2.MD.A.2

11. Which is the best estimate for the length of a flower bed?

A. 6 miles   **B.** 6 inches   **C.** 6 yards   **D.** 6 centimeters

2.MD.A.3

prepaze

MEASUREMENT & DATA

MEASURING LENGTH

12. Which is the best estimate for the height of a swing set?

   A. 9 feet   **B.** 9 inches   **C.** 9 miles   **D.** 9 centimeters

   2.MD.A.3

13. Which is the best estimate for the length of a basketball court?

   A. 30 feet   **B.** 30 yards   **C.** 30 miles   **D.** 30 inches

   2.MD.A.3

14. In Jack's bedroom, the poster of a dolphin is 3 feet tall. His soccer poster is 1 foot taller. How tall is the soccer poster?

   A. 4 feet   **B.** 2 feet   **C.** 3 feet   **D.** 31 feet

   2.MD.A.4

15. Cassie's paintbrush is 7 inches long. Alex's paintbrush is an inch shorter than Cassie's. How long is Alex's paintbrush?

   A. 7 inches   **B.** 8 inches   **C.** 6 inches   **D.** 5 inches

   2.MD.A.4

16. On David's street, the beige house is 25 feet tall. The blue house is a foot taller than the beige one. The red house is 2 feet shorter than the beige house. Which house is the tallest?

   A. Beige house   **B.** Blue house   **C.** Red house   **D.** All are same height

   2.MD.A.4

17. Daniel and Susan wanted to see how far they could jump. Daniel jumped 36 inches and Susan jumped 28 inches. How much farther did Daniel jump than Susan?

   A. 8 inches   **B.** 5 inches   **C.** 6 inches   **D.** 9 inches

   2.MD.A.5

MEASUREMENT & DATA

18. True or False: A whale is 78 feet long. A rhinoceros is 13 feet long. The difference in length between the lengths of the whale and the rhinoceros is 65 feet.

    A. True   **B.** False

(2.MD.A.5)

19. A building is 60 meters tall. A tree near the building is 12 meters tall. True or False: The building is 34 meters taller than the tree.

    A. True   **B.** False

(2.MD.A.5)

20. Billy had 1,000 grams of chocolate. If he ate 300 grams, how much chocolate did he have left?

Billy had _____ grams of chocolate left.

(2.MD.A.5)

MEASURING LENGTH

UNIT 2: TIME AND MONEY

prepaze

MEASUREMENT & DATA

TIME AND MONEY

1. The clock shows the time music class starts.

Music class is 40 minutes long. Which clock shows the time music class ends?

A.   **B.**   **C.**   **D.**

2.MD.C.7

2. Bill is riding the bus from his house to his grandmother's house. This clock shows the time he arrives.

The bus ride is 15 minutes long. Where would the hour and minute hands be on a clock at the time Bill leaves his house?

A. The hour hand is on the 8, and the minute hand on the 1.

B. The hour hand is between the 10 and 11, and the minute hand is on the 10.

C. The hour hand is on the 11, and the minute hand is on the 4.

D. The hour hand is between the 11 and 12, and the minute hand is on the 10.

2.MD.C.7

MEASUREMENT & DATA

3. Ivy rides her bike for 30 minutes. This clock shows the time she stops riding her bike.

What time did Ivy start riding her bike?

A. Eleven ten   **B.** Eleven forty

C. Ten thirty   **D.** Ten forty

(2.MD.C.7)

4. This clock shows the time Kai leaves her school on a bus.

The bus ride home is 20 minutes long. The walk from the bus stop to Kai's house is 5 minutes. Which clock shows the time Kai arrives home?

A.   3:30 PM   B.   4:10 PM

C.   3:25 PM   D.   3:15 PM

(2.MD.C.7)

prepaze

MEASUREMENT & DATA

TIME AND MONEY

5.   Sai takes 35 minutes to bake one pan of muffins. He is baking 2 pans of muffins, and he starts baking at 9:15 a.m.

Draw the time Sai finishes baking.

(2.MD.C.7)

6.   Justin leaves his friend's house at 2:55 pm. It takes him 10 minutes to walk home. Draw the time Justin arrives home on both clocks.

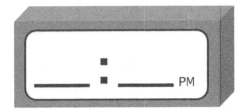

(2.MD.C.7)

MEASUREMENT & DATA

7. This clock shows the time Nat arrives at his friend's house.

Forty-five minutes later, he returns home. When he arrives home, he takes 35 minutes to eat dinner. What time does Nat finish dinner?

_____

2.MD.C.7

8. Lara watches TV for 1 hour. This clock shows the time she starts watching TV.

Which clock shows the time she stops watching TV?

A. 1:15 PM

B. 2:10 PM

C. 2:00 PM

D. 1:00 PM

2.MD.C.7

TIME AND MONEY

MEASUREMENT & DATA

TIME AND MONEY

9. John works on homework for 10 minutes. When he finishes, it is half past six. Draw the time John finishes his homework on these clocks.

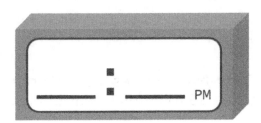

2.MD.C.7

10. How many minutes have passed since 8:00?

Explain your reasoning.

_____

_____

_____

2.MD.C.7

11. This is the money Zola has saved.

Zola's father gives her 58 more cents. How much money does Zola have now?

A. 22¢   **B.** 80¢

C. 36¢   **D.** 75¢

2.MD.C.8

MEASUREMENT & DATA

12. Umar has 4 dimes and 3 nickels. Lola has 2 quarters and 2 dimes.

Which skip counting strategy can be used to count value of these coins?

A. 25, 50, 55, 60, 65, 70, 75, 80, 85, 90, 95

B. 25, 50, 60, 70, 80, 90, 100, 110, 120, 130, 140

C. 25, 50, 60, 70, 80, 90, 100, 110, 115, 120, 125

D. 1, 2, 3, 4, 5, 6, 7, 8, 9, 10, 11

2.MD.C.8

13. Joey has 7 dimes and 8 nickels. Kari has 4 more dimes and 3 fewer nickels than Joey.

How much money do Joey and Kari have together?

Joey and Kari _____

2.MD.C.8

14. This is the school lunch menu.

Lunch Menu							
Item	Milk	Chicken Nuggets	Salad	Hotdog	Fries	Water	Juice
Cost	75¢	$2.40	$1.65	$1.00	75¢	55¢	95¢

Miles has 2 one-dollar bills, 3 quarters, 5 dimes, and 4 nickels. Which menu items is he able to buy and have the least money leftover?

A. Two hotdogs, fries, and a juice   **B.** A salad and chicken nuggets

C. A hotdog, salad, and a juice   **D.** Chicken Nuggets and fries

2.MD.C.8

prepaze

MEASUREMENT & DATA

TIME AND MONEY

15. Hyun has 5 one-dollar bills and 10 dimes. Jin has 4 one-dollar bills and 9 quarters. Who has more money, Hyun or Jin?

_____

(2.MD.C.8)

16. Lexi puts 4 one-dollar bills and 3 pennies in her bank. She now has $10.70 in the bank. How much money did Lexi start with?

A. $6.40   **B.** $14.73   **C.** $6.67   **D.** $6.73

(2.MD.C.8)

17. Diego has 65 cents. Hugo has 10 cents more than Diego. Jada has 25 cents more than Hugo. How much money do they have altogether?

A. $1.00   **B.** $2.40   **C.** $2.30   **D.** $1.50

(2.MD.C.8)

18. Sanjay has 4 quarters, 3 dimes, 8 nickels, and 2 pennies. He spends half of his money on candy. How much money does Sanjay have left?

_____

(2.MD.C.8)

19. Chin has 4 quarters, 3 dimes, 2 nickels and 11 pennies. Dara has 7 quarters, 5 dimes, 1 nickel, and 10 pennies. How could you determine who has more money? Explain your reasoning.

_____

_____

_____

(2.MD.C.8)

MEASUREMENT & DATA

20. Fredo has a one-dollar bill. He wants to give $0.35 to his friend. How can Fredo exchange his one-dollar bill for other coins so he can give money to his friend? Explain your reasoning.

_____

_____

_____

How much money would Fredo have left?

_____

2.MD.C.8

TIME AND MONEY

UNIT 3:   LEARN TO READ DATA

prepaze

NAME: .... DATE: ....

MEASUREMENT & DATA

1. A zookeeper measures a kitten's height.

= 3 inches

2 months

4 months

5 months

The kitten's height is measured from the top of the shoulders to the bottom of the feet. Which list shows the correct heights of the kitten?

A. 2 inches, 4 inches, 5 inches

B. 15 inches, 21 inches, 24 inches

C. 12 inches, 18 inches, 21 inches

D. 4 inches, 6 inches, 7 inches

2.MD.D.9

2. Alan creates a line plot to display his measurements.

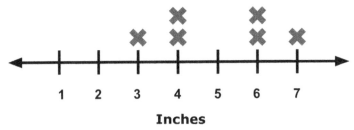

1   2   3   4   5   6   7

Inches

Which objects could Alan have measured?

A. Pencils   **B.** Chairs   **C.** Doors   **D.** Books

2.MD.D.9

MEASUREMENT & DATA

3. Measure each object to the nearest inch.

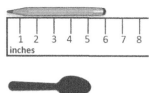

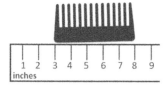

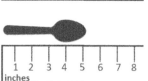

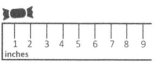

Which line plot matches these measurements?

A.

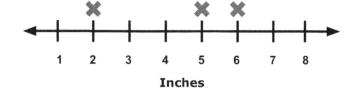

B.

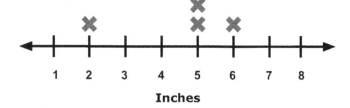

C.

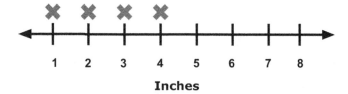

D.

2.MD.D.9

prepaze

MEASUREMENT & DATA

4. Which statement correctly describes the length of these pencils?

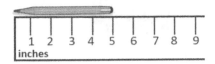
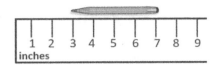
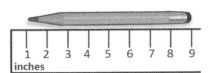
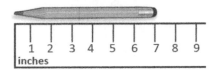

A. The shortest pencil is 4 inches long, and the longest pencil is 9 inches long.

B. The lengths of the pencils are from 4 to 8 inches.

C. Two of the pencils are 7 inches long.

D. The difference between the shortest and the longest pencil is 12 inches.

2.MD.D.9

5. Anvi creates a line plot to display her measurements.

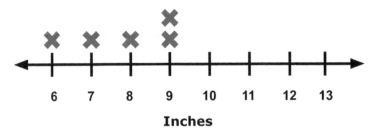

Which objects could Anvi have measured?

A. Scissors   **B.** Desks   **C.** Computers   **D.** Chairs

2.MD.D.9

MEASUREMENT & DATA

6. Daniel collects data on the lengths of different snakes.

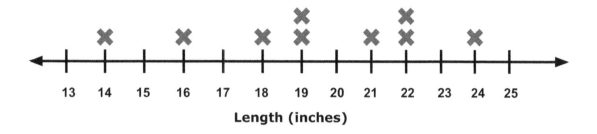

Length (inches)

What is the difference between the longest snake and the shortest snake?

_____ inches

2.MD.D.9

7. Use the data from this table to create a line plot below.

Length (inches)	Number of Pencils
5	II
7	III
8	IIII
9	II

How many pencils are 8 inches or longer?

_____

2.MD.D.9

prepaze

NAME: .... DATE: ....

MEASUREMENT & DATA

8. Asha uses a ruler to measure the length of these 5 leaves.

A.
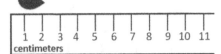

B.
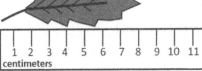

C.
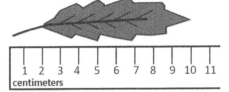

D.
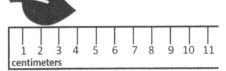

E.

Write the length of each leaf in the table.

Leaf	A	B	C	D	E
Length (centimeters)					

2.MD.D.9

prepaze

www.prepaze.com

MEASUREMENT & DATA

9. Ada measures the weight of her frog every week. This table shows her data.

Week	Weight (grams)
1	18
2	19
3	20
4	20

Which line plot represents this data?

A.

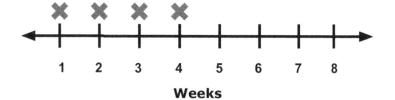

B.

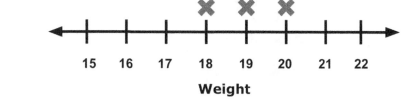

C.

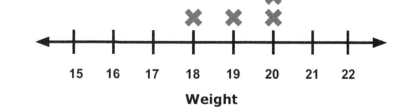

D.

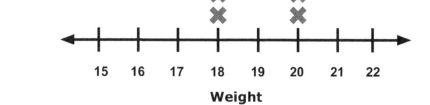

2.MD.D.9

MEASUREMENT & DATA

10. Use the data from this table to create a line plot in the space below.

Pencil Length (inches)	Number of Pencils
4	2
5	3
7	2
8	2
9	4
10	1

How many data points are on this line plot? _____

2.MD.D.9

11. This picture graph shows the number of ice cream cones sold.

Number of Ice Cream Cones Sold

Monday
Tuesday
Wednesday
Thursday

 means 2 ice cream cones

How many ice cream cones were sold altogether? _____

2.MD.D.10

MEASUREMENT & DATA

12. This picture graph shows the number of apples collected each day.

Number of Apples Collected

Monday	
Tuesday	
Wednesday	
Thursday	
Friday	

Each 🍎 means 2 apples

The number of apples collected on Friday is 5 more than the number of apples collected on Wednesday. How many apples are collected on Friday?

A. 5   **B.** 10

C. 9   **D.** 14

2.MD.D.10

13. This bar graph shows the insects Kyle saw while hiking.

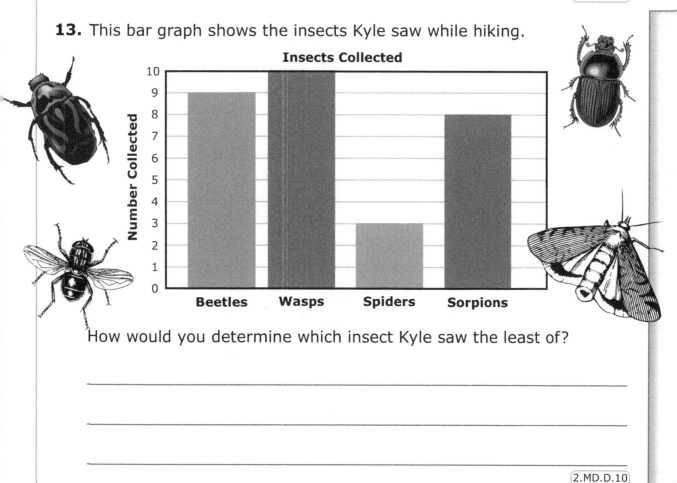

How would you determine which insect Kyle saw the least of?

_____

_____

_____

2.MD.D.10

prepaze

MEASUREMENT & DATA

14. How many more people like baseball than basketball?

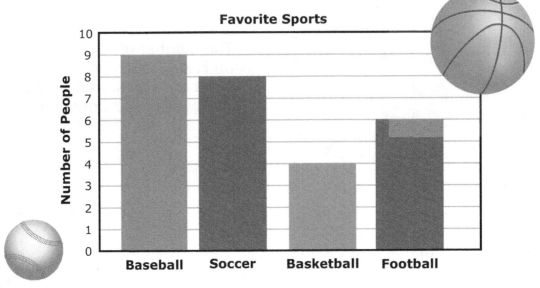

A. 13   **B.** 4   **C.** 9   **D.** 5

2.MD.D.10

15. This picture graph shows the number of miles 4 students in second-grade have to travel to get to school each day.

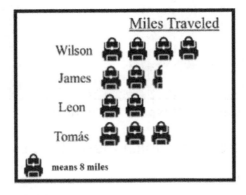

How many miles do the students travel altogether?

_____

2.MD.D.10

MEASUREMENT & DATA

16. Create a bar graph using this data.

- ▫ The number of cardinals is 3 more than the number of sparrows.
- ▫ There are 5 fewer sparrows than robins.
- ▫ There are 7 robins.
- ▫ There are 2 less robins than bluebirds.

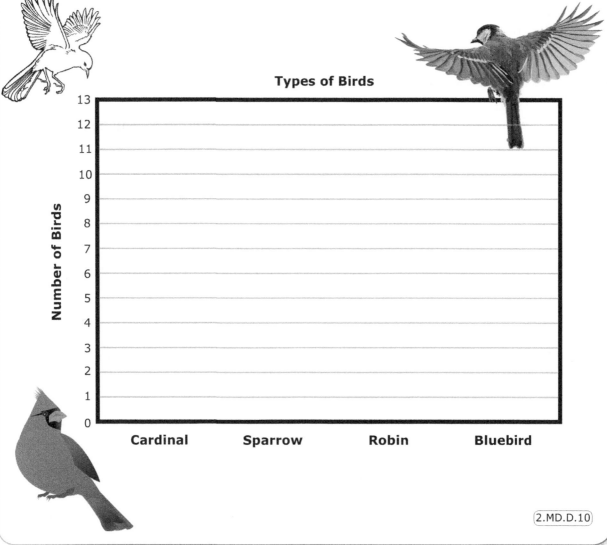

Types of Birds

Number of Birds (y-axis: 0 to 13)

Categories (x-axis): Cardinal, Sparrow, Robin, Bluebird

2.MD.D.10

prepaze

MEASUREMENT & DATA

17. This picture graph shows the number of miles an owl flies each day. From Monday to Thursday, the owl flies 84 miles. Complete the picture graph to show how many miles the owl flies on Wednesday.

Miles Traveled

Monday 🦉🦉🦉

Tuesday 🦉🦉🦉🦉🦉

Wednesday

Thursday 🦉🦉🦉🦉🦉🦉🦉

🦉 means 3 miles

How many owls do you have to add for Wednesday?

_____

_____

_____

(2.MD.D.10)

18. This picture graph shows the number of people who eat each type of cake.

Favorite Types of Cake

Strawberry

Chocolate

Vanilla

 means 3 people

Tommy believes 3 more people eat vanilla cake than chocolate cake. Do you agree? Explain your reasoning.

_____

_____

_____

(2.MD.D.10)

MEASUREMENT & DATA

19. This bar graph shows the favorite colors of the students in Mrs. Ryan's class.

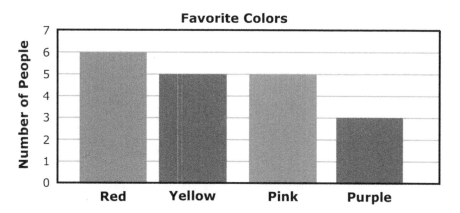

How many more students like red and yellow than pink and purple?

A. 8 **B.** 19 **C.** 3 **D.** 11

2.MD.D.10

20. Lissa is making this picture graph for different types of trees by her house.

Trees By Liam's House

Maple 🌲 🌲 🌲

Oak 🌲 🌲

Birch

Pecan

🌲 means 4 trees

There are 10 birch trees and 8 pecan trees. How would you draw the number of birch and pecan trees in this picture graph?

_____

_____

_____

_____

2.MD.D.10

CHAPTER REVIEW

prepaze

MEASUREMENT & DATA

1. How many more people are needed to like football to make it the favorite sport?

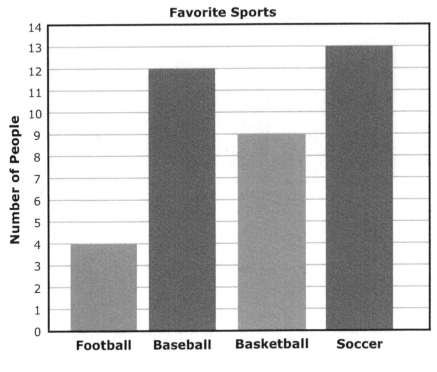

Favorite Sports

A. 4   **B.** 10   **C.** 13   **D.** 9

(2.MD.D.10)

2. This pictograph shows the number of ice cream cones sold.

Number of Ice Cream Cones Sold

Monday

Tuesday

Wednesday

Thursday

🍦 means 2 ice cream cones

How many more ice cream cones were sold on Wednesday than on Tuesday?

_____

(2.MD.D.10)

MEASUREMENT & DATA

3. Mr. Wilson is measuring the height of these 3 dogs.

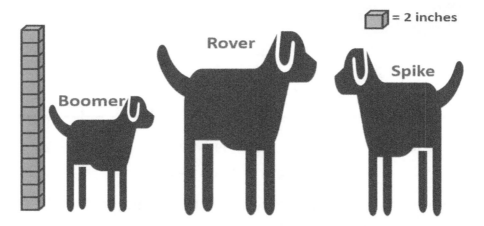

= 2 inches

He measures the dogs' height from the top of their shoulders to the bottom of their feet. Which table matches his measurements?

A.

Dog	Boomer	Rover	Spike
Height (inches)	14	22	20

B.

Dog	Boomer	Rover	Spike
Height (inches)	20	28	24

C.

Dog	Boomer	Rover	Spike
Height (inches)	7	11	10

D.

Dog	Boomer	Rover	Spike
Height (inches)	10	14	12

2.MD.D.9

prepaze

MEASUREMENT & DATA

4. David creates a line plot of the distances jumped by people in a contest.

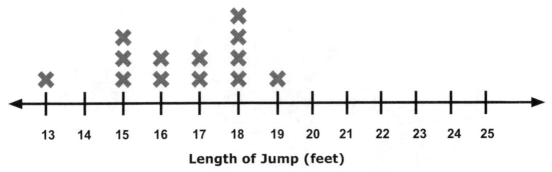

Length of Jump (feet)

How many more people jumped a distance longer than 15 feet than those who jumped 15 feet?

_____

2.MD.D.9

5. These are Zane's coins.

Zane gives forty-five cents to his sister. Which coins could you take away from this group to determine how much money Zane has left?

A. 1 quarter. 1 dime. and 2 nickels

B. 4 dimes and 1 nickel

C. 1 quarter and 4 dimes

D. 3 dimes. 3 nickels. and 5 pennies

2.MD.C.8

MEASUREMENT & DATA

6. Ellen has $1.75. She then receives 5 dimes and 4 pennies from her sister. Ellen's brother gives her twice as much money as her sister gave her. How much money does Ellen have now?

A. $2.38   **B.** $2.29   **C.** $2.04   **D.** $3.37

(2.MD.C.8)

7. Which digital clock matches the time shown on this analog clock?

A. `7:05 PM`   **C.** `12:07 PM`

B. `12:35 PM`   **D.** `12:30 PM`

(2.MD.C.7)

8. Bomi uses this strategy to determine the time shown on the clock.

> *First, I know that the hour hand is between 11 and 12, which means it is after 11 o'clock.*
>
> *Next, I count by fives, 9 times.*
>
> 5, 10, 15, 20, 25, 30, 35, 40, 45.
>
> *The time showing on the clock is 11:45.*

Do you agree with Bomi? Explain your reasoning.

_____

_____

(2.MD.C.7)

MEASUREMENT & DATA

9. Meg and Ryan are measuring the lengths of two pencils using this number line.

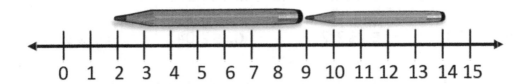

0 1 2 3 4 5 6 7 8 9 10 11 12 13 14 15

What is the combined length of both pencils on this number line?

A. 14 units **B.** 12 units **C.** 23 units **D.** 15 units

2.MD.B.6

10. Draw a line segment on this number line to represent a length of 7 units.

15   20   25   30   35   40

2.MD.B.6

11. Lin and Heidi compared their jump ropes. Lin's jump rope is 58 inches long and Heidi's jump rope is 70 inches long. How much longer is Heidi's jump rope than Lin's jump rope?

A. 14 inches **B.** 12 inches **C.** 2 inches **D.** 13 inches

2.MD.B.5

12. I have 3 pencils. The pencils are 17 cm, 12 cm, and 9 cm long.

True or False: The total length of all my pencils is 38 cm.

A. True **B.** False

2.MD.B.5

MEASUREMENT & DATA

13. Mary's boots are 8 inches tall. Jeff's boots are 2 inches taller. How tall are Jeff's boots?

A. 16 inches   **B.** 4 inches   **C.** 6 inches   **D.** 10 inches

2.MD.A.4

14. Caroline's bedroom is 14 long. Jameson's bedroom is 2 feet longer. True or False: Jameson has the smaller bedroom.

A. True   **B.** False

2.MD.A.4

15. Which is the best estimate for the length of a cherry stem?

A. 2 yards   **B.** 2 inches   **C.** 2 miles   **D.** 3 feet

2.MD.A.3

16. Which is the best estimate for the height of a bucket of sand?

A. 1 inch   **B.** 1 millimeter   **C.** 1 mile   **D.** 1 foot

2.MD.A.3

17. True or False: The screw is 6 buttons long or 5 dice long.

A. True   **B.** False

2.MD.A.2

prepaze

MEASUREMENT & DATA

18. How many paper clips long is the shoe? _____

How many buttons long is the shoe? _____

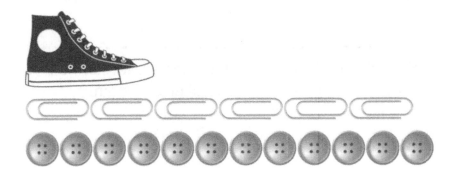

2.MD.A.2

19. True or False: The car measures 5 units in length.

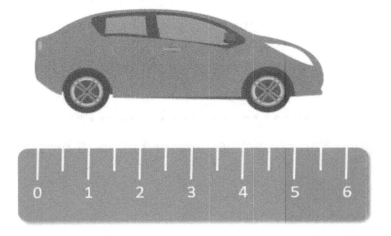

A. True

B. False

2.MD.A.1

MEASUREMENT & DATA

20. Which tool would you use to find out how long you slept during a nap?

 **Tape measure**

A.

 **Measuring cup**

B.

 **Clock**

C.

 **Scale**

D.

(2.MD.A.1)

EXTRA PRACTICE

prepaze

MEASUREMENT & DATA

1. Which tool would be the best for finding out how long it takes to walk up the stairs?

A.  Tape measure

B. Scale

C. Stopwatch

D. Calendar

2.MD.A.1

2. Which tool would you use to find out the height of a doorway?

A. Scale

B. Thermometer

C. Tape measure

D. Calendar

2.MD.A.1

3. How many paper clips long is the lizard? _____

How many dice long is the lizard? _____

2.MD.A.2

MEASUREMENT & DATA

4. How would you describe the length of the ice cream using paper clips and dice?

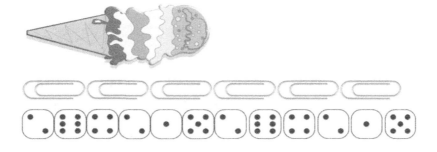

_____

_____

_____

(2.MD.A.2)

5. Which is the best estimate for the length of a bus?

A. 14 miles   **B.** 14 feet   **C.** 14 yards   **D.** 14 inches

(2.MD.A.3)

6. **True or False:** Two meters is a good estimate for the height of a basketball player.

A. True   **B.** False

(2.MD.A.3)

7. Paul grew 2 inches taller while he was in sixth grade. He was 5 feet 3 inches tall in the beginning of sixth grade. How tall is Paul now?

_____

(2.MD.A.4)

MEASUREMENT & DATA

8. The elm tree in my front yard is 3 meters tall and the red maple is 2 meters taller. How tall is the red maple?

_____

2.MD.A.4

9. The length of Joanne's loaf of Cuban bread is 56 centimeters. She cut off 32 centimeters of bread. What is the length of what she has left?

_____ centimeters

2.MD.B.5

10. The height of Hugo's math book is 19 centimeters. His reading book is 15 centimeters taller. What is the height of his reading book?

_____ centimeters

2.MD.B.5

11. Von is putting cylinder blocks into a box. The box is the width of 2 of these cylinders.

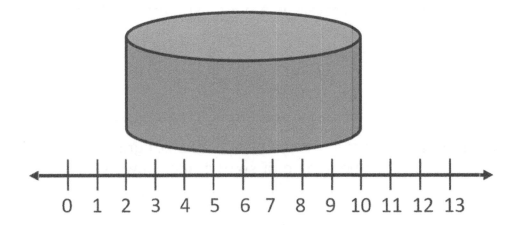

...question 11. continued next page

MEASUREMENT & DATA

Which number line models the width of the box?

A.

B.

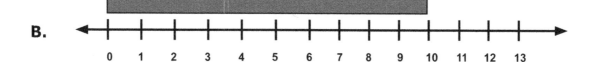

C.

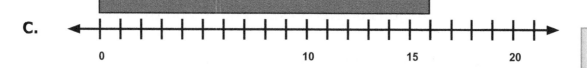

D.

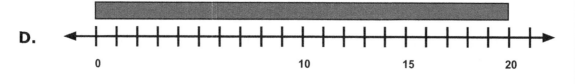

2.MD.B.6

12. Donna cuts a piece of string into 4 equal size pieces. This is one piece of the string.

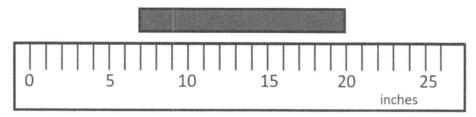

How long is the entire string?

A. 20 inches   **B.** 13 inches   **C.** 52 inches   **D.** 80 inches

2.MD.B.6

prepaze

MEASUREMENT & DATA

13. This clock shows the time school ends.

After school, Joni spends 10 minutes walking home. Where would the hour and minute hands be on the clock when Joni arrives home?

A. The hour hand is on the 3, and the minute hand is on the 12.

B. The hour hand is on the 2, and the minute hand is on the 12.

C. The hour hand is between the 2 and 3, and the minute hand is on the 8.

D. The hour hand is on the 4, and the minute hand is on the 12.

(2.MD.C.7)

14. Kari is feeding her dog and cat. This clock shows the time Kari starts feeding both animals.

$$2:15 \text{ PM}$$

Kari feeds her dog first. The dog takes 45 minutes to eat his food.

Kari's cat takes 15 minutes less time to eat than her dog takes.

What time does Kari's cat finish eating? Explain your reasoning.

_____

_____

_____

(2.MD.C.7)

MEASUREMENT & DATA

15. Josie has 5 quarters and 1 nickel. Liam has 3 dimes, 2 nickels, and 10 pennies. Evie has 1 dollar and 5 dimes. Dev has 6 quarters, 4 dimes, and 8 pennies.

Which list shows a comparison of the amount of money each one has, in order from greatest to least?

A. Liam, Dev, Josie, Evie

B. Dev, Evie, Josie, Liam

C. Dev, Josie, Liam, Evie

D. Liam, Josie, Evie, Dev

2.MD.C.8

16. On Monday, Ken put this amount of money in his bank.

He put the same amount of money in his bank on Tuesday, Wednesday and Thursday. How much money did Ken put in the bank?

2.MD.C.8

prepaze

MEASUREMENT & DATA

17. This line plot shows the distance some students live from their school.

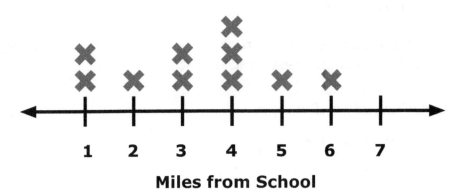

Miles from School

Which statement(s) about the data shown on this line plot are true? Explain your reasoning.

1: There are 5 students who live less than 3 miles from school.

2: Most of the students live more than 2 miles from school.

3: There are more students who live 3 miles from school, than students who live 2 miles from school.

_____

_____

_____

_____

_____

(2.MD.D.9)

MEASUREMENT & DATA

18. Ethan wants to make a line plot to show the temperature in different cities this month. How could he use this bar graph to create the line plot?

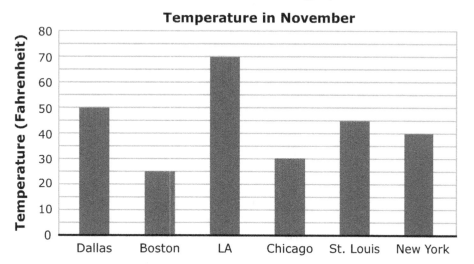

Temperature in November

_____

_____

_____

(2.MD.D.9)

19. This bar graph shows the votes for favorite colors of the students in Mr. Plum's class.

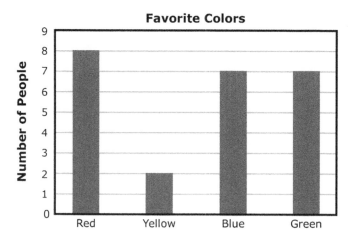

What is the difference between the number of votes for the most favorite color and the number of votes for the least favorite color?

A. 5   **B.** 6

C. 1   **D.** 8

(2.MD.D.10)

MEASUREMENT & DATA

20. Hyun makes a list of his friends' favorite animals.

Name	Animal
Bob	lion
Clara	monkey
Haley	tiger
Felipe	tiger
Fang	giraffe
Fredo	lion
Ian	lion
Dara	lion
Daniel	monkey
Jada	tiger
Kai	giraffe
Leo	monkey

Explain how you would create a bar graph to represent this data.

_____

_____

_____

2.MD.D.10

GEOMETRY

www.prepaze.com

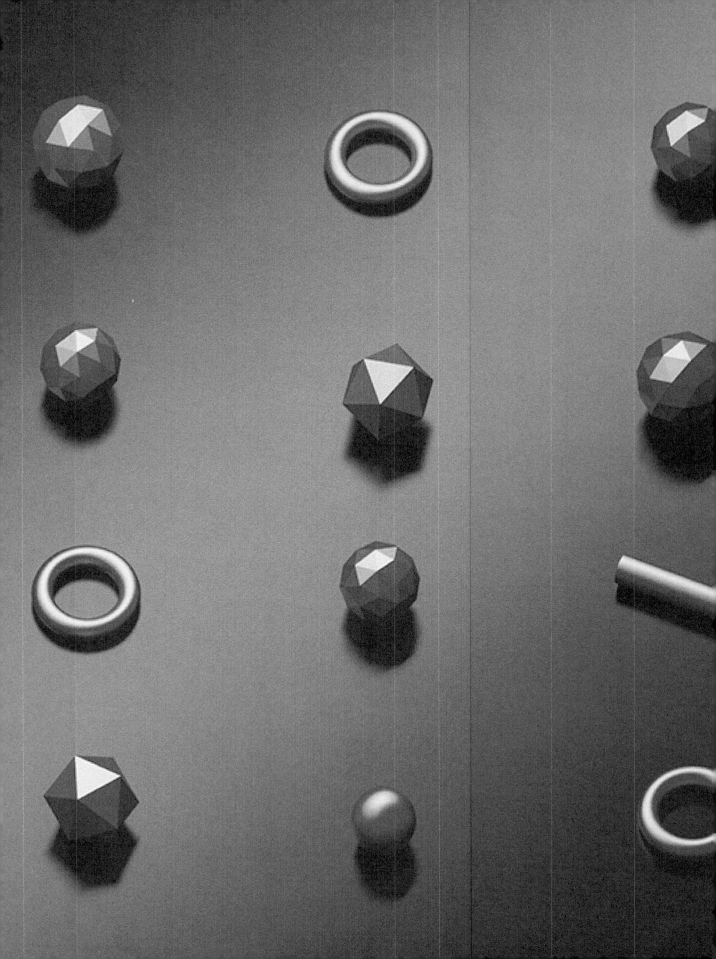

GEOMETRY

1. How many shapes are in both circles? _____

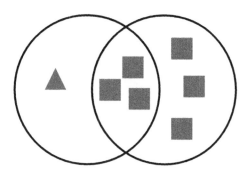

2.G.A.1

2. How many shapes are squares? _____

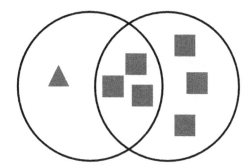

2.G.A.1

3. How many shapes are squares in the right circle but not in the left circle? _____

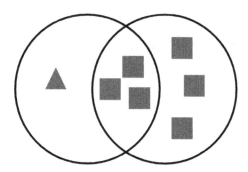

2.G.A.1

NAME: .... DATE: ....

GEOMETRY

UNDERSTANDING SHAPES

4. How many shapes are in the left circle? _____

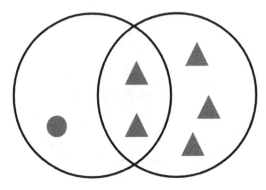

2.G.A.1

5. How many shapes are circles? _____

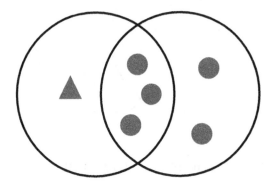

2.G.A.1

6. How many sides does the following shape have?

  Sides: _____

2.G.A.1

GEOMETRY

For questions 7–9, circle the 3-D shape that you can trace to make the 2-D shape in each row.

7.

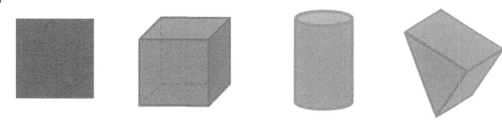

8.

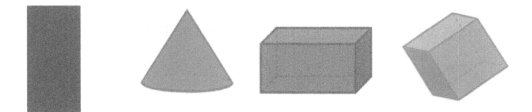

9.

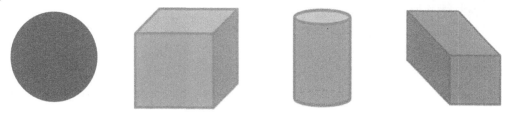

2.G.A.1

10. How many vertices does the following shape have?

 **Vertices:** _____

2.G.A.1

prepaze

NAME: .... DATE: ....

GEOMETRY

11. How many sides does the following shape have?

Sides: _____

2.G.A.1

12. How many vertices does the following shape have?

Vertices: _____

2.G.A.1

13. How many sides does the following shape have?

Sides: _____

2.G.A.1

GEOMETRY

14. How many vertices does the following shape have?

Vertices: _____

2.G.A.1

15. How many sides does the following shape have?

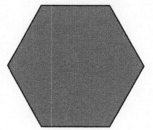

Sides: _____

2.G.A.1

16. How many vertices does the following shape have?

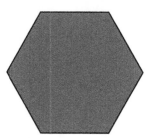

Vertices: _____

2.G.A.1

prepaze

GEOMETRY

UNDERSTANDING SHAPES

For questions 17–20, draw a line matching the shape with the real-life objects.

17.    

18.    

19.

20.

2.G.A.1

UNIT 2: PARTITION SHAPES INTO ROWS AND COLUMNS

GEOMETRY

1. Mya cut this rectangle into 2 rows and 3 columns.

Which number sentence shows how many pieces of paper Mya has?

A. 3 + 3 + 3   **B.** 2 + 2 + 2

C. 2 + 3   **D.** 2 + 2

2.G.A.2

2. Lara has a pan of brownies. She cuts the brownies into equal size pieces. This picture shows how many brownies are in the pan.

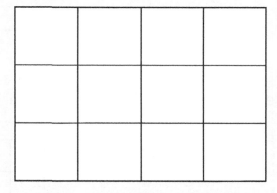

How many rows and columns does Lara create?

A. 4 rows and 3 columns   **B.** 1 row and 4 columns

C. 3 rows and 4 columns   **D.** 1 row and 12 columns

2.G.A.2

prepaze

GEOMETRY

PARTITION SHAPES INTO ROWS AND COLUMNS

3. Jon wants to divide a rectangular chocolate bar into 8 equal size pieces. How many columns and rows could he create?

A. 2 rows and 2 columns   **B.** 4 rows and 4 columns

C. 6 rows and 2 columns   **D.** 4 rows and 2 columns

2.G.A.2

4. Which array shows 3 rows of 5?

A.

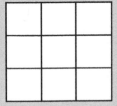

B.

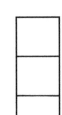

C.

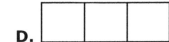

D.

2.G.A.2

5. Which response describes this array?

A. 3 rows of 2 squares

B. 3 rows of 3 squares

C. 9 rows of 1 squares

D. 1 row of 9 squares

2.G.A.2

6. Which response describes this array?

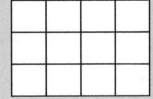

A. 3 rows of 4 squares

B. 3 rows of 3 squares

C. 12 rows of 1 square

D. 4 rows of 3 squares

2.G.A.2

prepaze

GEOMETRY

7. Draw an array with 3 rows of 6 squares.

(2.G.A.2)

8. Draw an array with 2 rows of 2 squares.

(2.G.A.2)

9. Emily draws an array with 4 rows of 3 squares. Mark draws an array with 4 rows of 3 squares. Whose array has more squares?

_____

(2.G.A.2)

10. Gino draws an array with 5 rows of 2 squares. Ivan draws an array with 4 rows of 4 squares. Whose array has more squares?

_____

(2.G.A.2)

prepaze

GEOMETRY

PARTITION SHAPES INTO ROWS AND COLUMNS

11. Draw an array with 4 columns and 5 rows of squares.

(2.G.A.2)

12. Draw an array with 3 columns and 6 rows of squares.

(2.G.A.2)

13. Draw an array inside the rectangle to match this equation:

$$2 + 2 + 2 + 2 = 8$$

(2.G.A.2)

GEOMETRY

14. Draw an array inside the rectangle to match this equation:

$$3 + 3 + 3 + 3 + 3 = 15$$

2.G.A.2

15. Create an array without any gaps or overlaps using this square. The array should have 4 rows and 4 columns.

2.G.A.2

prepaze

GEOMETRY

16. Create an array without any gaps or overlaps using this square. The array should have 3 rows and 1 column.

2.G.A.2

---

17. This rectangle is partitioned into 1 column and 3 rows to create smaller squares. How many smaller squares are created by this partition?

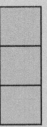

A. 3 square units

B. 2 square units

C. 5 square units

D. 4 square units

2.G.A.2

---

18. This square is partitioned into 2 columns and 2 rows to create smaller squares. How many smaller squares are created by this partition?

A. 4 square units

B. 2 square units

C. 6 square units

D. 1 square unit

2.G.A.2

GEOMETRY

19. This rectangle is partitioned into 6 columns and 2 rows to create smaller squares. How many smaller squares are created by this partition?

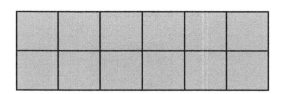

    A. 10 square units

    B. 8 square units

    C. 12 square units

    D. 4 square units

(2.G.A.2)

20. This rectangle is partitioned into 2 columns and 3 rows to create smaller squares. How many smaller squares are created by this partition?

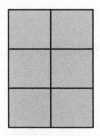

    A. 3 square units

    B. 8 square units

    C. 6 square units

    D. 5 square units

(2.G.A.2)

UNIT 3: PARTITION SHAPES INTO EQUAL PARTS

GEOMETRY

PARTITION SHAPES INTO EQUAL PARTS

1. Select all the pictures that show equal parts.

A.

B.

C.

D.

2.G.A.3

2. Select all the pictures that show equal parts.

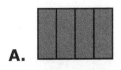
A.

B.

C.

D.

2.G.A.3

3. Select all the pictures that show equal parts.

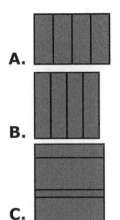
A.

B.

C.

D.

2.G.A.3

4. Select all the pictures that show equal parts.

A.

B.

C.

D.

2.G.A.3

GEOMETRY

5. Select all the pictures that show equal parts.

A.   **B.**   **C.**   **D.**

2.G.A.3

---

6. Which figure shows halves?

A.

B.

2.G.A.3

7. Which figure shows fourths?

A.

B.

2.G.A.3

---

8. Which figure shows halves?

A.

B.

2.G.A.3

9. Which figure shows fourths?

A.

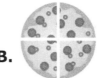

B.

2.G.A.3

prepaze

GEOMETRY

10. Which figure shows halves?

A.

B.

2.G.A.3

11. Which name describes how this rectangle is partitioned?

A. Halves
B. Thirds
C. Fourths
D. Fifths

2.G.A.3

12. Which sentence describes the shaded part of this square?

A. A third of the square is shaded.
B. Half of the square is shaded.
C. A fourth of the square is shaded.
D. A second of the square is shaded.

2.G.A.3

13. Which sentence describes the shaded part of this circle?

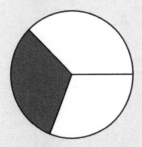

A. Half of the circle is shaded.
B. A third of the circle is shaded.
C. A fourth of the circle is shaded.
D. A fifth of the circle is shaded.

2.G.A.3

GEOMETRY

14. Which name describes the parts in this rectangle?

A. Halves   **B.** Thirds

C. Fourths   **D.** Fifths

2.G.A.3

15. Which name describes the parts in this rectangle?

A. Halves   **B.** Thirds

C. Fourths   **D.** Fifths

2.G.A.3

16. Which shape shows fourths?

A.   **B.**   **C.**   **D.**

2.G.A.3

17. Which shape shows two halves?

A.   **B.**   **C.**   **D.**

2.G.A.3

prepaze

GEOMETRY

PARTITION SHAPES INTO EQUAL PARTS

18. Which shape shows three thirds?

 A.   **B.**   **C.**   **D.**

2.G.A.3

19. Which picture shows a fourth of a shape is shaded?

 A.   **B.**   **C.**   **D.**

2.G.A.3

20. Which shape shows four fourths?

 **A.**   **B.**    **C.**    **D.**

2.G.A.3

CHAPTER REVIEW ➡

prepaze

GEOMETRY

1. Carl draws a shape with more than 3 vertices and less than 6 sides. Which shape did Carl draw?

A. 　　B. 　　C. 　　D.

2.G.A.1

2. Jin has 12 tomato seeds. She wants to plant the seeds in a large container with 12 equal parts. Which container should Jin use?

A. 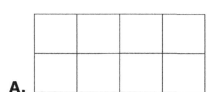　　B.

C.　　D.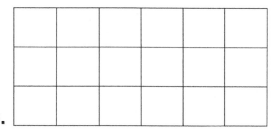

2.G.A.2

3. Which picture shows halves?

A. 　　B. 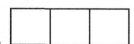　　C. 　　D.

2.G.A.3

GEOMETRY

4. Which statement does not describe a triangle?

    A. A triangle has more sides than a rectangle.

    B. A triangle is a shape with 3 angles.

    C. A triangle has fewer sides than a pentagon.

    D. A triangle is a shape with 3 sides.

    2.G.A.1

5. Tam cuts this cake into pieces.

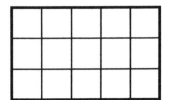

How many pieces of cake does she cut?

    A. 15   **B.** 12   **C.** 10   **D.** 13

    2.G.A.2

6. Which phrase describes this model?

    A. Three halves   **B.** Three ones

    C. Three thirds   **D.** Three threes

    2.G.A.3

7. How many faces are on a cube?

    A. 4   **B.** 5   **C.** 6   **D.** 8

    2.G.A.1

8. Jennifer divides a rectangle into 4 rows and 2 columns to create same-size squares. How many same-size square parts does the rectangle have?

    A. 8   **B.** 6   **C.** 10   **D.** 12

    2.G.A.2

GEOMETRY

9. Jon divides a pizza into 4 equal shares. Which name describes the shares he created?

A. Fours   **B.** Fourteens   **C.** Fourths   **D.** Fourteenths

(2.G.A.3)

10. Which response describes quadrilaterals?

A. A quadrilateral has fewer sides than a pentagon.

B. A quadrilateral has only 3 edges.

C. A quadrilateral has 6 vertices.

D. A quadrilateral has more sides than a hexagon.

(2.G.A.1)

11. Darren cuts a rectangular ice cream sandwich into 6 equal sized pieces. Which sentence describes how Darren cuts the ice cream sandwich?

A. Darren cuts the sandwich into 3 rows and 3 columns.

B. Darren cuts the sandwich into 6 rows and 1 column.

C. Darren cuts the sandwich into 5 rows and 1 column.

D. Darren cuts the sandwich into 2 rows and 2 columns.

(2.G.A.2)

12. Show four fourths in this rectangle.

(2.G.A.3)

prepaze

GEOMETRY

13. Kwan says these shapes all have 6 faces. Is Kwan correct? Explain your reasoning.

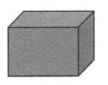 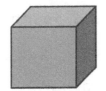 

_____

_____

_____

2.G.A.1

14. Michael wants to divide a rectangular chocolate bar into 12 equal size pieces. How many columns and rows could he create?

A. 8 rows and 4 columns   **B.** 12 rows and 2 columns

C. 6 rows and 6 column   **D.** 4 rows and 3 columns

2.G.A.2

15. Show three-thirds inside this circle.

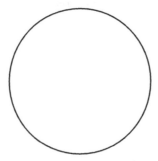

2.G.A.3

GEOMETRY

16. Draw a shape matching this description:

- ▫ The shape has 4 angles.
- ▫ The shape has 4 sides.
- ▫ Two sides with different lengths.

2.G.A.1

17. Complete the number sentence to describe this array.

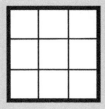

_____ + _____ + _____ = _____

2.G.A.2

18. How would you describe the shaded part of this circle?

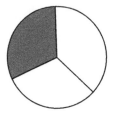

_____

_____

_____

2.G.A.3

GEOMETRY

19. Ken says a pentagon has more vertices than a hexagon. Is this true? Explain why or why not.

_____

_____

_____

2.G.A.1

20. Uke draws this array.

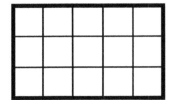

Write a number sentence to represent this array.

_____

2.G.A.3

EXTRA PRACTICE

GEOMETRY

1. Linh draws a shape with 12 edges and 6 faces. Which shape does Linh draw?

A. Octagon   **B.** Hexagon   **C.** Cube   **D.** Quadrilateral

2.G.A.1

2. Geno wants to divide this rectangle into 14 smaller squares.

Which response describes how he can divide the rectangle?

A. Divide the rectangle into 2 rows and 7 columns

B. Divide the rectangle into 4 rows and 10 columns

C. Divide the rectangle into 5 rows and 9 columns

D. Divide the rectangle into 14 rows and 14 columns

2.G.A.2

3. Jerry shades half of one-third of a rectangle. Which drawing shows the part of the rectangle Jerry shades?

A.   **B.**   **C.**   **D.**

2.G.A.3

prepaze

GEOMETRY

4. Liam creates 2 shapes using 9 toothpicks. Which two shapes can Liam create?

A. Pentagon and Hexagon   **B.** Hexagon and Square

C. Pentagon and Cube   **D.** Triangle and Hexagon

2.G.A.1

5. Which array has the same number of squares as an array made of 3 rows of 4 squares?

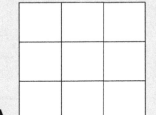
A.

B.

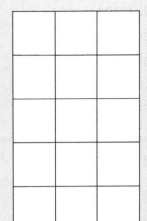
C.

D.

2.G.A.2

prepaze

GEOMETRY

6. Which sentence describes the shaded part of this rectangle?

    A. A third of the rectangle is shaded.

    B. Half of the rectangle is shaded.

    C. A fourth of the rectangle is shaded.

    D. A second of the rectangle is shaded.

2.G.A.3

7. Which statement about rectangles is true?

    A. A rectangle has four corners, which means it has eight sides.

    B. A rectangle has four corners, which means it has four angles.

    C. A rectangle has four sides, which means it has eight vertices.

    D. A rectangle has six faces, which means it has eight vertices.

2.G.A.1

8. Which statement describes this array?

    A. 3 rows of 2 squares   **B.** 3 rows of 3 squares

    C. 9 rows of 1 square   **D.** 1 row of 9 squares

2.G.A.2

prepaze

NAME: .... DATE: ....

GEOMETRY

9. Which shape is divided into 4 equal shares?

A.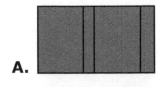

B.

C.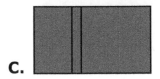

D.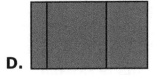

`2.G.A.3`

10. List 2 ways these shapes are alike.

_____

_____

_____

`2.G.A.1`

11. Use this square to create an array showing 1 row of 4 squares.

`2.G.A.2`

prepaze

www.prepaze.com

GEOMETRY

12. Which is larger: three thirds or two halves? Explain your reasoning.

_____

_____

_____

(2.G.A.3)

13. Which of these shapes are pentagons?

A.   **B.**   **C.**

(2.G.A.1)

14. Hal draws an array with 2 rows of 6 squares. Eric draws an array with 5 rows of 3 squares. Whose array has more squares?

_____

(2.G.A.2)

15. Shade 2 halves of this circle.

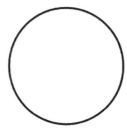

(2.G.A.3)

prepaze

GEOMETRY

EXTRA PRACTICE

16. Which shape has the most sides?

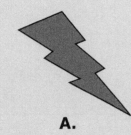

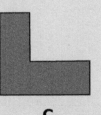

A.   B.   C.

2.G.A.1

17. Draw an array with 6 columns and 3 rows of squares.

2.G.A.2

18. Zoe says these rectangles both show 4 fourths. Do you agree?
Explain your reasoning.

   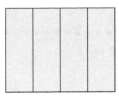

2.G.A.3

GEOMETRY

19. Which shape is a hexagon?

 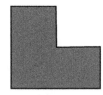

A.   B.

2.G.A.1

20. Complete the array using square tiles inside the rectangle to match this equation:

$$2 + 2 + 2 + 2 + 2 = 10$$

2.G.A.2

COMPREHENSIVE ASSESSMENTS

prepaze

ASSESSMENT ①

COMPREHENSIVE ASSESSMENTS

1. A farmer needs to bring 90 vegetables to the market. He picked 53 tomatoes and 31 cucumbers. How many more vegetables does he need for the market?

He needs _____ more vegetables.

(2.OA.A.1)

2. Samantha is writing a repeated addition equation to go with her picture. How many times does she need to add the number 2?

Samantha needs to add the number 2 _____ times.

(2.OA.C.4)

3. I am thinking of a three-digit number that has 5 ones, 7 hundreds, and 0 tens. What number am I thinking of? How do you know?

_____

_____

_____

(2.NBT.A.1)

prepaze

COMPREHENSIVE ASSESSMENTS

4. Jeremy has 530 pennies in his piggy bank. He finds 10 more pennies. How many pennies does he have now? How do you know?

_____

_____

_____

2.NBT.A.2

5. Which response describes this array?

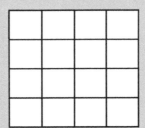

A. 4 rows of 4 squares
B. 8 rows of 8 squares
C. 16 rows of 1 square
D. 1 row of 16 squares

2.G.A.2

6. Count the parts in the whole. Which name describes the parts?

A. Three halves
B. Two halves
C. Two ones
D. Three ones

2.G.A.3

COMPREHENSIVE ASSESSMENTS

7. The expression $800 + 30 + 7$ is the expanded form of what number? How do you know?

_____

_____

_____

_____

(2.NBT.A.3)

ASSESSMENT 1

8. Ryan says to compare the numbers 683 and 724, you should look at the digit in the tens place. Do you agree or disagree? Why?

_____

_____

_____

_____

(2.NBT.A.4)

9. Abraham read 6 chapters of his book. He has 9 more chapters to read. How many chapters are in his book?

A. 15

C. 27

B. 24

D. 20

(2.OA.B.2)

prepaze

COMPREHENSIVE ASSESSMENTS

10. Sky wrote the equation $7 + 7 = 14$ to describe this picture. Is she correct? Why or why not?

_____

_____

_____

2.OA.C.3

11. The rule for this table is "subtract 106".

Fill in the blanks.

In	Out
574	_____
203	_____
701	_____
1,000	_____

2.NBT.B.7

12. *Use mental math.* Mr. Owens spends $225 over five days. He spends $100 on Monday, and $100 on Tuesday. How much money does he spend on the other days?

$_____

2.NBT.B.8

COMPREHENSIVE ASSESSMENTS

13. AJ asks the students in his school to name their favorite color.

- ▫ 39 people like yellow
- ▫ 55 people like blue
- ▫ 82 people like green
- ▫ 71 people like red
- ▫ 43 people like purple

What strategy could AJ use to find the total number of students who like red, yellow, or blue?

A. $(71 + 39) + 55$ **B.** $39 + 55 + 82$

C. $(40 + 71) + 1 + (55)$ **D.** $(39 + 70 + 55) - 1$

2.NBT.B.9

14. What is the measure of the length of the line, to the nearest centimeter?

A. 2 centimeters   **B.** 1 centimeter

C. 4 centimeters   **D.** 6 centimeters

2.MD.A.1

15. How many dice or buttons are needed to measure the line?

A. 1 dice or 1 button

B. 2 dice or 1 button

C. 3 dice or 2 buttons

D. 4 dice or 1 button

2.MD.A.2

COMPREHENSIVE ASSESSMENTS

16. Which is the best estimate for the length of a moving truck?

A. 17 yards   **B.** 17 inches   **C.** 17 miles   **D.** 17 centimeters

2.MD.A.3

17. This balance measures the mass of an apple in grams.

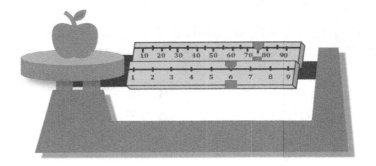

What is the mass of this apple? _____ grams

2.NBT.B.5

18. This number line shows how Zara adds 56, 11 and 13.

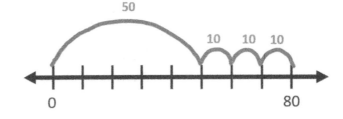

Do you agree with her strategy? Explain your reasoning.

_____

_____

_____

2.NBT.B.6

COMPREHENSIVE ASSESSMENTS

19. Tom's sailboat is 14 feet long. Andrea's sailboat is 2 feet longer. Who has the longest sailboat?

A. Tom   **B.** Andrea

(2.MD.A.4)

20. Taru uses this strategy to add 800 and 194.

> *First 1 split all the numbers and added the hundreds, tens and ones to get 8 hundreds, 9 tens, and 4 ones.*
>
> *The number is 891.*

Do you agree with his strategy? Why or why not?

_____

_____

_____

(2.NBT.B.7)

21. Nia uses the length of this block to model a number.

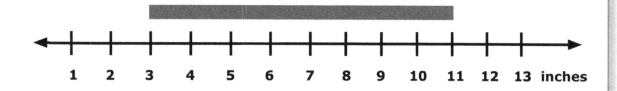

| 1 | 2 | 3 | 4 | 5 | 6 | 7 | 8 | 9 | 10 | 11 | 12 | 13 | inches |

What is the length of this block?

A. 13 inches   **B.** 9 inches   **C.** 8 inches   **D.** 11 inches

(2.MD.B.6)

ASSESSMENT 1

prepaze

COMPREHENSIVE ASSESSMENTS

22. What arithmetic expression is modeled on this number line? Explain your reasoning.

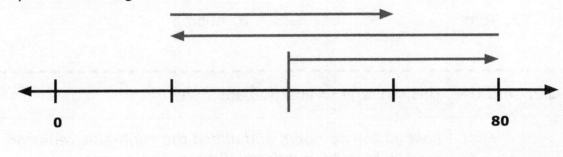

0   80

_____

_____

_____

2.NBT.B.5

23. Kevin spends $75 at the store. Melissa buys these items at the store.

$25 $31 $29

Who spends more money, Kevin or Melissa? Explain your reasoning.

_____

_____

_____

2.NBT.B.6

COMPREHENSIVE ASSESSMENTS

24. Which digital clock matches the time shown on the analog clock?

A. 11:50 PM

B. 10:50 PM

C. 10:10 PM

D. 10:55 PM

2.MD.C.7

25. One box is 42 cm tall. Another box is 18 cm tall. How tall will the boxes be if you stack them on top of each other?

A. 70 cm   **B.** 60 cm   **C.** 24 cm   **D.** 50 cm

2.MD.B.5

26. This graph shows the number of animals on a farm.

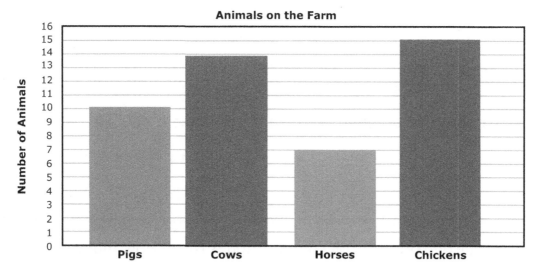

The farmer buys 5 more animals. How many animals are on the farm now?

A. 41   **B.** 51   **C.** 46   **D.** 40

2.MD.D.10

prepaze

COMPREHENSIVE ASSESSMENTS

27. Which shape is 2 dimensional?

    A. triangle   **B.** triangular prism   **C.** cube   **D.** sphere

2.G.A.1

28. Lee uses this strategy to add 17,48 and 29.

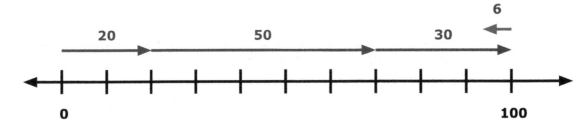

Do you agree with his strategy? Explain your reasoning.

_____

_____

_____

2.NBT.B.6

29. What number can go in both blanks to make a doubles fact?

_____ + _____ = 14

2.OA.C.3

30. Meg put her coins into 5 columns and 3 rows. How many coins does Meg have?

Meg has _____ coins.

2.OA.C.4

COMPREHENSIVE ASSESSMENTS

31. How many hundreds are in the number 673?

A. 0 **B.** 3 **C.** 6 **D.** 7

(2.NBT.A.1)

32. Maya is counting by 5's what should she say after 985?

A. 980 **B.** 990 **C.** 995 **D.** 1000

(2.NBT.A.2)

33. Jake's family has driven 693 miles. Write this number in word form.

_____

(2.NBT.A.3)

34. This is the school lunch menu.

Lunch Menu					
Item	Milk	Hamburger	Salad	Fries	Water	Juice
Cost	85¢	$2.65	$1.50	85¢	50¢	75¢

Pari has 4 quarters, 6 dimes, and 3 nickels. Which menu items is she able to buy and have the least amount of money left over?

A. Fries and milk **B.** Salad and juice
C. Hamburger and fries **D.** Salad and water

(2.MD.C.8)

35. Mina is trying to write the following comparison. "Nine hundred eighty-one is greater than eight hundred twenty-two."

Help her write the comparison using numbers and < or > symbols.

_____

(2.NBT.A.4)

prepaze

COMPREHENSIVE ASSESSMENTS

36. Measure each object to the nearest inch.

Which line plot matches these measurements?

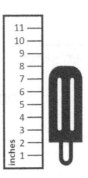

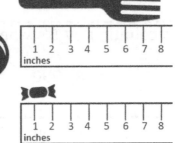

A.

1   2   3   4   5   6   7   8

Inches

B.

1   2   3   4   5   6   7   8

Inches

C.

2   3   4   5   6   7   8   9

Inches

D.

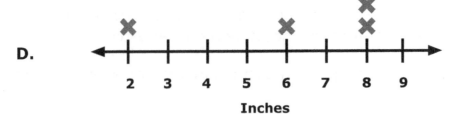

2   3   4   5   6   7   8   9

Inches

2.MD.D.9

COMPREHENSIVE ASSESSMENTS

37. Theo has a total of 400 marbles in 3 different bags.

175 marbles   86 marbles   ?

Write an addition number sentence to find the number of marbles in the question mark bag. _____

How many marbles are in the question mark bag? _____

2.NBT.B.7

38. Johann weighs 80 kilograms, Andre weighs 70 kilograms. Bennie weighs 90 kilograms.

True or False: Together, all three boys weigh 240 kilograms.

A. True   **B.** False

2.MD.B.5

39. Vani goes to bed at 8:30 pm. Before getting into bed, she reads a book for 35 minutes. She plays with her dog for 10 minutes before she reads her book.

Draw the time when Vani starts playing with her dog on this clock.

2.MD.C.7

COMPREHENSIVE ASSESSMENTS

40. Three students draw a line segment to model 16 on a number line. Which number lines are correct? Explain your reasoning.

A.

B.
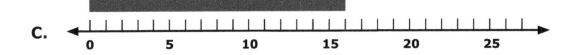

C.
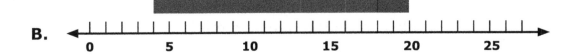

_____

_____

_____

2.MD.B.6

41. Raley has 8 one-dollar bills and 2 dollars in coins. She gives the money shown below to her friend Lara.

Raley gives half of the money she has left to her sister. How much money does Raley have left?

_____

2.MD.C.8

COMPREHENSIVE ASSESSMENTS

42. *Use mental math.* Each hash mark is 200. Start at Point A on the number line.

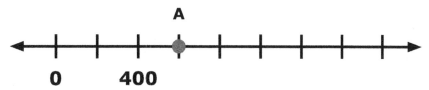

Subtract 100, and then subtract 10. What number are you on?

_____

2.NBT.B.8

43. Which strategy can be used to add these numbers?

$$79 + 25 + 11 + 25 - 100$$

A. $100 - (79 + 25 + 11 + 11)$ **B.** $(179 + 50 + 11)$
C. $(79 + 11) - 150$ **D.** $(79 + 11) + (25 + 25) - 100$

2.NBT.B.9

44. This line plot shows the number of miles a bird flies each day.

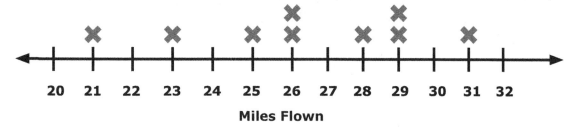

Miles Flown

How would you determine the total number of days the bird flies?

_____

_____

_____

2.MD.D.9

prepaze

COMPREHENSIVE ASSESSMENTS

ASSESSMENT 1

45. Jin is planting flowers in the school garden. This bar graph shows the number of flowers he plants.

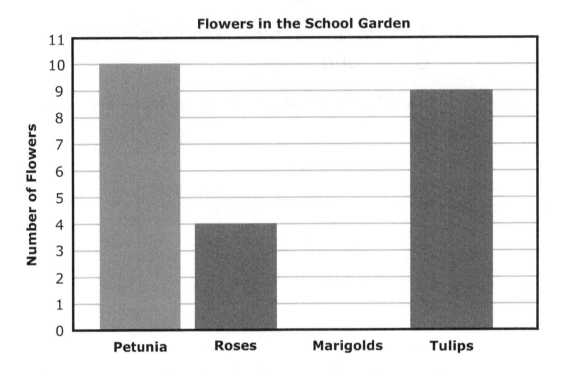

Jim wants to plant marigolds, if he plants 6 more marigolds than roses. How many marigolds will he plant?

_____

2.MD.D.10

ASSESSMENT ②

COMPREHENSIVE ASSESSMENTS

ASSESSMENT ②

1. Matt's Flower Shop is packing flowers for an order. They packed 32 yellow flowers, 4 red flowers and 6 bundles of orange flowers. Each bundle has 10 flowers in it. How many flowers have they packed?

They packed _____ flowers.

2.OA.A.1

2. Jacob put his books into 5 columns and 5 rows. How many books does Jacob have?

He has _____ books.

2.OA.C.4

3. Write the number that is equal to 2 hundreds and 5 tens.

_____

2.NBT.A.1

4. Maria has 250 marbles. Sophia has 220 more marbles. How many marbles does Sophia have?

A. 48   **B.** 480   **C.** 400   **D.** 470

2.NBT.A.2

5. Which response describes this array?

A. 3 rows of 5 squares   **B.** 15 rows of 3 squares
C. 5 rows of 3 squares   **D.** 1 row of 15 squares

2.G.A.2

COMPREHENSIVE ASSESSMENTS

6. Count the parts in the whole. Which name describes each part?

A. Halves
B. Thirds
C. Fourths
D. Fifths

2.G.A.3

7. Jessica has 931 pennies in her piggy bank. How can she write this number in expanded form?

_____

2.NBT.A.3

8. Which symbol goes in the blank to make the statement true?

8 hundreds and 7 tens _____ 807

A. <
B. >
C. =
D. !

2.NBT.A.4

9. There are 17 pencils in a box. Fourteen children take one pencil each. How many pencils are still in the box?

A. 4
B. 11
C. 6
D. 3

2.OA.B.2

10. A bakery has to sell the cupcakes shown below. Does the expression 9 + 9 represent the number of cupcakes the bakery has to sell? Explain your reasoning.

_____

_____

_____

2.OA.C.3

prepaze

COMPREHENSIVE ASSESSMENTS

11. Describe a strategy you would use to add these numbers.

$$474 + 135$$

_____

_____

_____

_____

2.NBT.B.7

12. *Use mental math.* Explain how you would find the missing numbers in this table.

Input	Output
?	512
635	645
740	750
890	?

_____

_____

_____

_____

2.NBT.B.8

COMPREHENSIVE ASSESSMENTS

13. Violet describe the strategy she uses to subtract these two numbers.

$$87 - 34$$

> *First, I round the numbers in the tens place to the nearest ten.*
>
> $$90 - 30 = 60$$
>
> *Next, I subtract the numbers in the ones place.*
>
> $$7 - 4 = 3$$
>
> *Then, I subtract my answers.*
>
> $$60 - 3 = 57$$
>
> *Last, since I rounded my first number up by 3, I have to subtract 3.*
>
> $$57 - 3 = 54$$

Do you agree with Violet's strategy? Explain your reasoning.

_____

_____

_____

(2.NBT.B.9)

14. What is the measure of the length of the line, to the nearest centimeter?

A. 2 centimeters   **B.** 6 centimeters

C. 3 centimeters   **D.** 4 centimeters

(2.MD.A.1)

prepaze

COMPREHENSIVE ASSESSMENTS

15. True or False: Including the tail, the dog is 4 paper clips or 5 buttons long.

A. True

B. False

(2.MD.A.2)

16. True or False: Twenty-six centimeters is a good estimate for the length of an ear of corn.

A. True   **B.** False

(2.MD.A.3)

17. The length of the pillow shown on this bed is 18 inches.

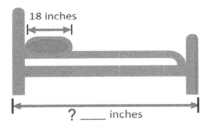

18 inches

? _____ inches

Explain how you would estimate the length of the entire bed, in inches.

_____

_____

_____

(2.NBT.B.5)

COMPREHENSIVE ASSESSMENTS

18. Lori, Vince, and Jack are playing a video game.

- ▫ Lori scores 15 points more than Vince.
- ▫ Vince scores 26 points.
- ▫ Jack scores 13 points more than Lori.

How many points does Jack score?

_____ points

2.NBT.B.6

19. Madison's ice cream cone is 9 inches tall. Jose's ice cream cone is 2 inches shorter than Madison's. Lana's ice cream cone is an inch longer than Madison's. Who has the shortest ice cream cone?

A. Madison **B.** Jose **C.** Lana

2.MD.A.4

20. Use the blocks to draw a model of this equation in the space below.

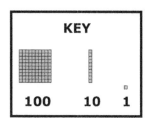

495 + 197 = _____

2.NBT.B.7

prepaze

COMPREHENSIVE ASSESSMENTS

21. Ben has 20 more pennies than Adam. Ben has 53 pennies. How could you use this number line to find the number of pennies Adam has? Explain your reasoning.

_____

_____

_____

2.MD.B.6

22. Explain how you would write a number sentence to represent the difference between Model A and Model B.

KEY

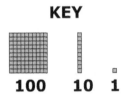

100   10   1

Model A

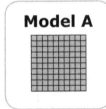

Model B

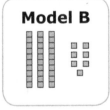

_____

_____

_____

2.NBT.B.5

prepaze

COMPREHENSIVE ASSESSMENTS

23. This number line models the distance a bird flies as it travels across the country.

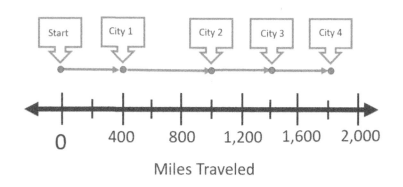

Write a number sentence to represent the total distance the bird travels.

_____

(2.NBT.B.6)

24. Luke starts his art project at a quarter to five. He spends 25 minutes making his art project.

Draw the time Luke finishes his art project.

(2.MD.C.7)

prepaze

ASSESSMENT ②

COMPREHENSIVE ASSESSMENTS

ASSESSMENT ②

25. John collects coins. This table shows the different types of coins he collects.

Coin	Pennies	Nickels	Dimes	Quarters
Number Collected	15	10	18	?

John makes this bar graph to model his data.

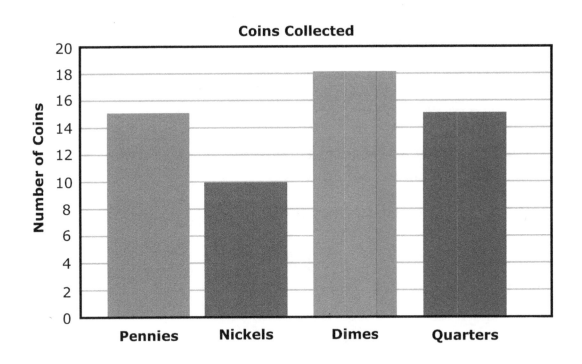

How many quarters did he collect? _____

2.MD.D.10

COMPREHENSIVE ASSESSMENTS

26. Yesterday I had 1,000 grams of chocolate. Brandon ate 500 grams of my chocolate and Allison ate 200 grams of my chocolate. How much chocolate do I have left?

_____ grams

2.MD.B.5

27. Which three-dimensional shapes have the same number of edges?

A.

B.

C.

D.

2.G.A.1

28. Yani, Edray, Wes, and Han are playing a video game.

- ▫ Yani scores 20 points less than Edray.
- ▫ Edray scores 13 points more than Wes.
- ▫ Wes scores 38 points.
- ▫ Han scores 19 points more than Yani.

How many points do they score altogether?

_____ points

2.NBT.B.6

29. There are _____ cars in each column.

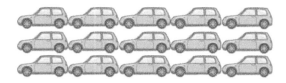

2.OA.C.4

prepaze

COMPREHENSIVE ASSESSMENTS

ASSESSMENT ②

30. Which doubles fact can be used to represent the total number of fish in the fish tank?

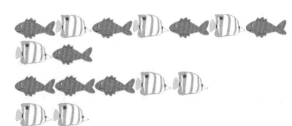

     A. 6 + 6

     B. 9 + 9

     C. 8 + 8

     D. 8 + 9

(2.OA.C.3)

31. Write the number that is equal to 0 tens 7 hundreds and 0 ones.

_____

(2.NBT.A.1)

32. Sam is counting by ones. What number does he say directly before 739?

     A. 738   **B.** 749   **C.** 750   **D.** 639

(2.NBT.A.2)

33. How many blocks are there? Write the number in expanded form.

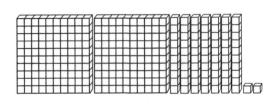

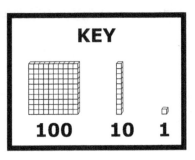

KEY

100   **10**   **1**

_____

(2.NBT.A.3)

COMPREHENSIVE ASSESSMENTS

34. Jane has 5 quarters, 4 dimes, 6 nickels, and 11 pennies. How do you know she has more than one dollar?

    A. One dollar has the same value as 10 dimes.

    B. Jane has more nickels than dimes.

    C. Jane has 11 pennies.

    D. Five quarters, alone, is greater than one dollar.

(2.MD.C.8)

35. Dev creates a line plot of the heights of the students in his class.

Height (inches)

How many more students are shorter than 51 inches than those who are taller than 51 inches?

_____

(2.MD.D.9)

36. Use the symbols <, >, or = to compare the values in the statement.

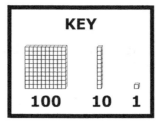

KEY

100   10   1

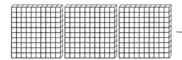 _____ 9 tens, 2 hundreds, and 5 ones

(2.NBT.A.4)

prepaze

COMPREHENSIVE ASSESSMENTS

37. Which point on the number line represents this sum?

$$145 + 87$$

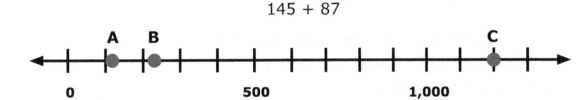

2.NBT.B.7

38. Rudolf is 130 cm tall. Hermann is 5 cm taller than Rudolf.

True or False: Hermann is 135 cm tall.

A. True   **B.** False

2.MD.B.5

39. The students in Mr. Burr's class are in the lunchroom and on the playground.

- There are 7 students on the playground.
- There are 9 more students in the lunchroom than on the playground.

Draw a line segment on this number line to represent the total number of students in Mr. Burr's class.

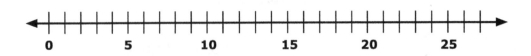

2.MD.B.6

prepaze   www.prepaze.com

COMPREHENSIVE ASSESSMENTS

40. Tommy uses this strategy to determine the time shown on the clock.

> *First, I know that half past 9 o'clock is 9:30. The minute hand would be on the 6.*
>
> *Since the minute hand is on the 7, I add one more minute.*
>
> *The time showing on the clock is 9:31 pm.*

Do you agree with Tommy? Explain your reasoning.

_____

_____

_____

2.MD.C.7

41. Liz has 9 coins and 3 one-dollar bills in her pocket. This picture shows some of the coins and bills.

Liz has a total of $3.96. What other coins and bills does she have inside her pocket? Explain your reasoning.

_____

_____

_____

2.MD.C.8

COMPREHENSIVE ASSESSMENTS

42. *Use mental math.*

- There are 125 students in the second grade.
- The number of students in third grade is 10 less than the number of students in fourth grade.
- The number of students in fourth grade is 10 more than the number of students in first grade.
- The number of students in first grade is 10 more than the number of students in second grade.

How many students are in third grade?

_____ students

2.NBT.B.8

43. Zaria describe the strategy she uses to subtract two numbers.

$$300 - 24$$

> *First, I start with the number 300. I don't have any ones or tens to subtract, so I must regroup.*
>
> *One hundred can be regrouped as 10 tens, but since I need ones to subtract, I'm going to regroup 1 ten as 10 ones.*
>
> *Now I have 2 hundreds 9 tens and 10 ones. Now I can subtract!*
>
> $$10 - 4 = 6$$
> $$90 - 20 = 70$$
>
> *I have 200 + 70 + 6 or 276.*

Do you agree with Zaria's strategy? Explain your reasoning.

_____

_____

_____

2.NBT.B.9

COMPREHENSIVE ASSESSMENTS

44. This picture graph shows the number of miles an owl flies each day.

Miles Traveled

On which day did the owl fly the

farthest? _____

2.MD.D.10

45. Evie wants to make a line plot to model the shoe length of different people in her class. How could she use this bar graph to create the line plot?

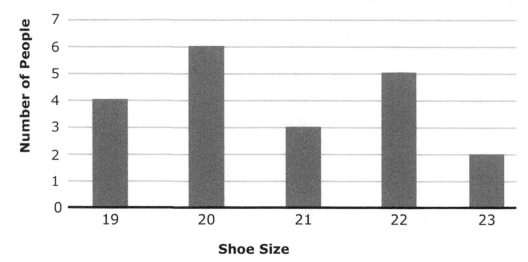

Shoe Length of People in my Class

_____

_____

_____

2.MD.D.9

prepaze

ANSWERS AND EXPLANATIONS

prepaze

ANSWERS and EXPLANATIONS

OPERATIONS AND ALGEBRAIC THINKING
UNIT 1: ADD AND SUBTRACT WITHIN 20

1 Answer: D
Explanation: This is a comparison situation. To find 13 more than 6, add: 13 + 6 = 19.

2 Answer: A
Explanation: This is a separate (result unknown) situation. Represent this situation with subtraction: 20 − 9 = 11.

3 Answer: C
Explanation: This is a separate (result unknown) situation. Represent this with subtraction: 18 − 11 = 7.

4 Answer: A
Explanation: This is a separate (result unknown) situation. Represent this situation with subtraction: 15 − 6 = 9.

5 Answer: B
Explanation: This is a joining (change unknown) situation. This could be represented as 11 + = 19 or 19 − 11 = 8.

6 Answer: B
Explanation: This is a separate (result unknown) situation. Represent this situation with subtraction: 14 − 8 = or 14 − 8 = 6.

7 Answer: C
Explanation: This is a part-part-whole problem. Represent this situation with addition: 9 + 7 = _____ or 9+7=16.

8 Answer: D
Explanation: This is a joining situation. Represent this situation with addition: 9 + 12 = _____ or 9+12=21.

9 Answer: D
Explanation: This is a comparison situation. Represent this situation with addition and then subtraction with with the following equation: (9+8)−13=4.

10 Answer: C
Explanation: This is a part-part-whole situation. Represent this situation with subtraction: 18−5 =_____ or 18−5=13.

11 Answer: B
Explanation: 13 + ? = 20 is equivalent to 20−13= ?.

12 Answer: D
Explanation: Find the number of wheels adding the number of wheels on one car 3 times. 4+4+4=12.

13 Answer: A
Explanation: A total of 9 books have been read so far. Use this equation: 18−(4+5)=18−9=9.

14 Answer: D
Explanation: This is a separation situation, represented using subtraction: 20 − 2 = _____ or 20−2=18.

15 Answer: B
Explanation: This is a separation situation, represented using subtraction: 17 − 13 = ____ or 17−13=4.

16 Answer: C
Explanation: This is a separation situation, represented using subtraction: 17 − 6 = _____ or 17−6=11.

17 Answer: B
Explanation: Find the answer using two equations: 13−1=12 and then 12+4=16.

18 Answer: A
Explanation: This is a joining situation, represented by addition: 8 + 12 = _____ or 8+12=20.

19 Answer: D
Explanation: This is a part-part- whole situation, represented using subtraction: 15 − 8 = _____ or 15−8=7.

20 Answer: B
Explanation: This is a part-part-whole situation, represented as 7 + 4 + _____ = 20 or 20−(7+4)=9.

OPERATIONS AND ALGEBRAIC THINKING
UNIT 2 : ADDITION AND SUBTRACTION
WORD PROBLEMS

1 Answer: C
Explanation: Add: 48 plus 48 equals 96.

2 Answer: C
Explanation: Add and then subtract: 43 plus 37 equals 80. 95 minus 80 equals 15.

3 Answer: D
Explanation: Add and then subtract: 13 plus 6 equals 19. 19 minus 4 equals 15.

4 Answer: D
Explanation: Add: 16 plus 16 plus 21 equals 53.

5 Answer: C
Explanation: Subtract: 84 minus 62 is 22.

6 Answer: B
Explanation: Mr. Frank started with 83 stickers. He gave away 24 and 17 so subtract both numbers from 83.

7 Answer: A
Explanation: Add: 26 plus 28 plus 23 equals 77.

8 Answer: 48
Explanation: Subtract: 90 minus 42 equals 48.

9 Answer: 39
Explanation: Subtract: 53 minus 14 equals 39.

10 Answer: 86
      +
Explanation: Subtract and then add: They sold 14 cars and 18 more cars were delivered.

11 Answer: 28
Explanation: Add: 28 plus 14 equals 42. Then, subtract: 70 minus 42 equals 28.

12 Answer: 92
Explanation: Add: 48 plus 16 plus 28 equals 92.

13 Answer: 90
Explanation: Add: 30 plus 30 plus 30 equals 90.

14 Answer: 85
Explanation: Add: 47 plus 38 equals 85.

15 Answer: 51
Explanation: Answer must include that Anthony needs 51 more shells and a reasonable explanation such as, "I know Anthony needs 51 more shells because 85 minus 34 equals 51."

16 Answer: 20
Explanation: Answer must include that there will be 20 lunches left and a reasonable explanation such as, "21 plus 33 equals 54 so 54 kids will eat lunch. There are 74 lunches minus the 54 that will be eaten leaves 20 extra lunches."

17 Answer: 90 − 62 = ?
          He has $28.
Explanation: Answer must include a reasonable equation and explanation such as, "I could use the equation 90 − 62 = ? To help me figure how much money Tom has left. If the answer is more than $20 left he could buy a show ticket. If it is less than $20 he could not buy a ticket."

18 Answer: Stephanie
Explanation: Answer must include that Stephanie brought more and a reasonable explanation such as, "15 plus 25 is 40 so Edward brought 40 items. Then, 24 plus 20 is 44 so Stephanie brought 44 items. The number 44 is more than 40 so Stephanie brought more treats to the bake sale."

19 Answer: 68
Explanation: Answer must include that she has read 68 pages so far and a reasonable

prepaze

explanation such as, "I subtracted 24 from 92. The answer was 68 so Jessica has read 68 pages so far."

20 Answer: 43
Explanation: Answer must include that there are 43 apples left and a reasonable explanation such as, "45 plus 29 equals 74 apples. Then, 31 apples are taken so subtract 31 and there are 43 apples left."

OPERATIONS AND ALGEBRAIC THINKING
UNIT 3: FOUNDATIONS FOR MULTIPLICATION

1 Answer: A
Explanation: There are 14 guitars, which is 7 sets of 2.

2 Answer: C
Explanation: There are 4 rows of 4, so 4 + 4 + 4 + 4 is 16.

3 Answer: C
Explanation: There are 16 cars, so divide 16 into two equal groups of 8.

4 Answer: B
Explanation: Columns are up and down, so each column has 3 objects.

5 Answer: D
Explanation: Move one monkey from the first row to the second row. Then, the doubles equation that matches the picture of 12 monkeys is 6 plus 6 equals 12.

6 Answer: A
Explanation: The picture shows 3 rows. Each row has 4 items in it, so add 4 + 4 + 4.

7 Answer: B
Explanation: There are two sets of 5 flowers with one extra flower.

8 Answer: 4
Explanation: There are 4 objects in each row, so he should add the number 4 three times, which gives 4 + 4 + 4.

9 Answer: odd
Explanation: Celeste has 15 stickers. 15 is an odd number.

10 Answer: 6
Explanation: The 3 rows of 2 is equivalent to 2 plus 2 plus 2 or 6.

11 Answer: 0
Explanation: The number 14 is an even number. You can put 7 on each table and have 0 left over.

12 Answer: 3
Explanation: There are 3 rows of 4 in each row, so the number 4 should be added 3 times: 4 + 4 + 4.

13 Answer: even
Explanation: There are 12 bikes, 12 is an even number.

14 Answer: 3 rows of dots with 3 dots in each row
Explanation: Answers must include a reasonable explanation such as, "I would draw three rows of dots. Each row would have 3 dots in it. Three rows of 3 dots each shows 3 plus 3 plus 3 equals 9."

15 Answer: 10
Explanation: There are 20 apples, so divide the 20 apples into two equal groups of 10.

16 Answer: There are 15 guitars; Explanations will vary
Explanation: Answers must include a reasonable explanation such as, "There are 5 guitars in each row and 3 rows of guitars. She can add 5 plus 5 plus 5 to get the total number of guitars." or "There are 3 guitars in each column and 5 columns of guitars. She can add 3 plus 3 plus 3 plus 3 plus 3 to get the total number of guitars."

**17 Answer: Even
14 is even.**
Explanation: There are 14 apples. The number 14 is even. Answers must include a reasonable explanation such as, "I counted

by 2's to figure out that there are 14 apples. Since I was able to count all the apples when I counted by 2's I know there is an even number of apples."

18   Answer: disagree; Explanations will vary

Explanation: Answers must include a reasonable explanation such as, "I disagree because the picture shows 3 rows of 5 not 4 rows of 5. His addition sentence should be 5 + 5 + 5."

19   Answer: agree; Explanations will vary

Explanation: Answers must include a reasonable explanation such as, "I agree because there are 4 pencils and you can count by 2's to get 4 so the number is even."

20   Answer: Odd;. Explanations will vary

Explanation: Answers must include a reasonable explanation such as, "There are an odd number of birds. I counted the birds by 2's and I had one bird left over so I know there are an odd number." or "There are 15 birds, and 15 is an odd number."

OPERATIONS AND ALGEBRAIC THINKING CHAPTER REVIEW

1   Answer: D
Explanation: This is a comparison situation represented using subtraction: $14 - 11 =$_____ or $14-11=3$.

2   Answer: C
Explanation: Add: 47 plus 32 equals 79. Then subtract: 79 minus 41 equals 38.

3   Answer: D
Explanation: Add: 10 plus 10 equals 20.

4   Answer: A
Explanation: Subtract: 87 minus 43 is 44, so there are 44 more cars than trucks.

5   Answer: A
Explanation: This is a two-step situation. Use these equations to find the number of crayons Daniel has left: $8+8=16$ and $16-4=12$.

6   Answer: B
Explanation: There are 12 fish which can be represented with the double fact 6 plus 6.

7   Answer: A
Explanation: This is a joining situation. Use the equation: $9+8=17$.

8   Answer: C
Explanation: There are 18 cherries, which can be represented with the double fact 9 plus 9.

9   Answer: C
Explanation: This is a separation situation which is represented using subtraction: $19-7=12$.

10   Answer: B
Explanation: The picture shows 4 rows. Each row has 5 items in it.

11   Answer: B
Explanation: This is a part-part-whole situation. Use addition: $16 + 9 = 25$.

12   Answer: C
Explanation: The picture shows 4 groups of 3. Add the number 3 four times.

13   Answer: 4
Explanation: Each row has 4 so she should start counting at 4.

14   Answer: 31; Explanation will vary
Explanation: Answer must include a reasonable explanation such as, "I know that there are 31 blocks left in the bin. I know because 94 minus 42 minus 21 equals 31."

15   Answer: 4
Explanation: There are 4 bikes in each row.

prepaze

16   Answer: 4;
$$37-(16 + 17) =?$$
Explanation: Answer must include a reasonable explanation such as, "I know there are four problems left. $37 - (16 + 17)$ = ?".

17   Answer: even
Explanation: There are 18 basketballs. 18 is an even number.

18   Answer: 9
Explanation: There are 18 stickers. Group 18 into two equal groups of 9.

19   Answer: 41
Explanation: Subtract: 92 minus 51 equals 41.

20   Answer: 3+3+3+3;
or 4+4+4;
Explanation will vary
Explanation: Answers must include a reasonable explanation such as, "The number sentence $3 + 3 + 3 + 3$ describes this picture. I know because I see 4 rows of apples. Each row has 3 apples in it."

OPERATIONS AND ALGEBRAIC THINKING EXTRA PRACTICE

1   Answer: D
Explanation: To find the total number of points add 45 and 41.

2   Answer: D
Explanation: This is a two-step problem. Step 1: Subtract 3 from 14. Step 2: Add 8 to the answer.$(14 - 3) + 8 = 19$.

3   Answer: A
Explanation: This is a joining situation. Add:$13 + 7 = 20$.

4   Answer: A
Explanation: This is a separation situation. Subtract: $19 -$ _____ $= 8$ or $19 - 8 = 11$

5   Answer: Count by 2s
Explanation: Answers must include a reasonable explanation such as, "Raymondo can count by 2s. If he can count all the beads by 2s the number is even. If he has one left over the number is odd."

6   Answer: Odd;
Explanations will vary
Explanation: Answers must include a reasonable explanation such as, "I know it's an odd number because I counted the lemons by 2s and had one left over."

7   Answer: B
Explanation: Among the choices, the only double fact is 5 plus 5.

8   Answer: D
Explanation: There are 15 strawberries. They can divide 15 into two groups of 7 with one left over.

9   Answer: D
Explanation: Each row has 5 items.

10   Answer: A
Explanation: 4 plus 4 plus 4 equals 12.

11   Answer: B
Explanation: He should draw 4 rows since the number 3 is added 4 times. He should put 3 in each row.

12   Answer: 36
Explanation: Subtract and then add: 73 minus 50 plus 13 equals 36.

13   Answer: 25
Explanation: 5 groups of 5 is 25 because 5 + 5 + 5 + 5 + 5 = 25.

14   Answer: 3
Explanation: She should add the number 4 three times since there are 3 rows.

15   Answer: 1
Explanation: There are 13 cherries. They can be grouped into two equal groups of 6 with 1 left over.

16 Answer: 10

Explanation: There are 20 stickers which can be grouped into two equal groups of 10.

17 Answer: Yes

Explanation: Answer must include a reasonable explanation such as, "Rachel has raised 79 dollars so far. I know because 48 plus 31 equals 79, and 79 is more than 75 so Rachel has exceeded her goal by $4."

18 Answer: Yes; Explanations will vary

Explanation: Answers must include a reasonable explanation such as, "I agree with Quan. When you count the cubes by 2s you get 8 sets of 2. So the double fact 8 plus 8 equals 16, represents the picture."

19 Answer: Disagree; Explanations will vary

Explanation: Answers must include a reasonable explanation such as, "I disagree because the chart has 2 columns and 4 rows."

20 Answer: 4 rows by 3 columns or 3 columns by 4 rows or 2 columns by 6 rows or 6 columns by 2 rows

Explanation: Answers must include a reasonable explanation such as, "She could put her 12 stickers into 3 rows with 4 stickers in each row." or "She could put her 12 stickers into 4 rows with 3 stickers in each row."

NUMBER AND OPERATIONS IN BASE TEN UNIT 1: UNDERSTANDING PLACE VALUE

1 Answer: B

Explanation: The tens place is the second digit. The digit 0 is in the tens place.

2 Answer: C

Explanation: 50 tens means 50 times 10 which is the same as 500.

3 Answer: A

Explanation: 8 hundreds, 7 tens, and 0 ones is 800 + 70 + 0 which is the number 870.

4 Answer: B

Explanation: 700 + 40 + 5 is the same as 745.

5 Answer: B

Explanation: When counting by 1s 640 comes after 639, and 639 + 1 = 640.

6 Answer: C

Explanation: 400 + 90 + 3 is the expanded form of four hundred ninety- three (493).

7 Answer: C

Explanation: When counting by 10s 900 comes after 890, and 890 + 10 = 900.

8 Answer: D

Explanation: The picture shows 4 hundreds blocks, 4 tens blocks and 4 ones blocks. 400 + 40 + 4 = 444.

9 Answer: A

Explanation: 917 is less than 970, so the symbol is <. This symbol always points to the smaller side.

10 Answer: B

Explanation: There are 4 hundreds blocks, 3 tens blocks, 4 ones blocks. 400 + 30 + 4 = 434.

11 Answer: C

Explanation: 845 is greater than 844, so the symbol is >. This symbol always points to the smaller side.

12 Answer: 100; Explanation will vary

Explanation: Answers must include a reasonable explanation such as, "Jessica has 100 cubes. 10 tens is the same as ten times ten which is one hundred. If you have 10 ten sticks of cubes you can exchange them for one hundred block."

13 Answer: 865; Explanation will vary

Explanation: Answers must include a reasonable explanation such as, "Mark should say 865 next. Counting by 5s means saying every 5th number. 865 is five more than 860."

14 Answer: 880

Explanation: 880 is the numeral form of eight hundred eighty.

15 Answer: Edwin

Explanation: Edwin has 879 and Scott has 789. 879 is greater than 789.

16 Answer: 993; Explanation will vary

Explanation: Answers must include a reasonable explanation such as, "Sean should say 993 next. Counting by 100s means adding one hundred. One hundred more than 893 is 993."

17 Answer: False

Explanation: 852 is not less than 843.

18 Answer: Agree; Explanation will vary

Explanation: Answers must include a reasonable explanation such as, "I agree. The picture shows 2 hundreds blocks, 4 tens blocks, and 2 ones blocks. That's 200 + 40 + 2"

19 Answer: 348

Explanation: 3 hundreds, 4 tens, and 8 ones is 300 + 40 + 8 and is 348.

20 Answer: 80

Explanation: There are 8 hundreds blocks; 800 is 80 times 10.

NUMBER AND OPERATIONS IN BASE TEN
UNIT 2: PROPERTIES OF OPERATIONS

1 Answer: A

Explanation: This is a separation (change unknown) situation, so the equation $37 - \underline{\hspace{1cm}} = 23$ represents this situation.

2 Answer: C

Explanation: The pattern in the graph increases by 3 each day. Day 6, Mr. Ono will travel 24 miles. The total miles will be 9 + 12 + 15 + 18 + 21 + 24 = 99.

3 Answer: B

Explanation: The distance between each hash mark on the number line is 20 units. The hash marks show $80 + 20 - 50$.

4 Answer: D

Explanation: The amount of money saved increases by $4 each week. The total will be 9 + 13 + 17 + 21 + 25 = 85.

5 Answer: 35

Explanation: Determine the number of books with the expression: $46 - 22 + 11$.

6 Answer: $16

Explanation: Find the total cost of all four items 30 + 18 + 12 + 24 = 84 and subtract 84 from 100.

7 Answer: 95

Explanation: Add 60 and 35. The other values / spaces on the game board are not relevant and are not added.

8 Answer: 7

Explanation: Combine the values given for reading, 25 + 18 and subtract 36 from the total: $43 - 36 = 7$.

9 Answer: 79

Explanation: The lines on the thermometer increase by 15 so the original temperature of the classroom was 75 degrees: $75 + 8 - 4 = 79$.

10 Answer: 72

Explanation: The goat travels up the mountain and back down the mountain each day. This is 8 miles each day, and 8 added 9 times is 72.

11 Answer: D

Explanation: Altogether, the team scored 62 points in Game 1, and 75 points in Game 2. Subtract: 75 – 62 = 13.

12 Answer: B

Explanation: Add the number of pages together: 17 + 16 +21 + 14 = 68.

13 Answer: D

Explanation: Josie exercised 45 minutes on Monday, 35, (15 + 20) minutes on Wednesday, 15 minutes on Friday. Altogether, she exercised 95 minutes: 45 + 35 + 15 = 95.

14 Answer: B

Explanation: This situation is a joining situation. Add the numbers: 15 + 15 + 12.

15 Answer: Hank and Mike

Explanation: Hank scored 21 points and Mike scored 19 points: 21 + 19 = 40.

16 Answer: watch, mittens, hat

Explanation: The hat cost 8 dollars, the watch cost 54 dollars, and gloves cost 10 dollars: 8 + 54 + 10 = 72.

17 Answer: 53

Explanation: Determine the age of John's father by adding: 15 + 28 + 10 = 53.

18 Answer: 25, 90

Explanation: Use the scale. Each icons used in the pictograph represents 10 students and each half icon represents 5 students.

19 Answer: 88

Explanation: Sam finishes in 23 minutes, Ronaldo finishes in 35 minutes. Together all three students finish in 88 minutes: 30 + 23 + 35 = 88.

20 Answer: 200 + 300 = 500

Explanation: The distance between City 2 and City 3 is 200 miles and the distance between City 3 and City 4 is 300 miles. So, 200 + 300 = 500 represents this situation.

NUMBER AND OPERATIONS IN BASE TEN
UNIT 3: PLACE VALUE TO ADD AND SUBTRACT

1 Answer: D

Explanation: Three hops of 25 is equal to 75. 55 plus 75 is 130.

2 Answer: A

Explanation: 3 hops of 150 is equal to 450. 110 plus 450 is 560.

3 Answer: 228
457
698
1,000

Explanation: Add 99 to each input ("in") value to get the output ("out") value.
129 + 99 = 228
358 + 99 = 457
599 + 99 = 698
901 + 99 = 1,000

4 Answer: 210

Explanation: Building #1 has 24 windows on each side, 24 + 24 +24 = 72. Building#2 has 36 windows on each side, 36 + 36 + 36 = 108. Building #3 has 10 windows on each side, 10 + 10 +10 = 30. The total number of windows needed is 72 + 108 + 30 = 210.

5 Answer: 233

Explanation: Plot 621 on the number line then jump left 388 to 233.

6 Answer: 89 + _____ = 462
5 – 1 = 4

Explanation: The correct number sentence is 89 + ? = 462. A variety of strategies could be used, including estimation, to determine how many flowers can be planted in the pot. A correct answer is 5 – 1 = 4.

prepaze

7　Answer: 597, represented with 5 hundreds blocks, 9 tens blocks, and 7 ones blocks.

Explanation:　135 is drawn using 1 hundreds block, 3 tens blocks, and 5 ones blocks. 462 is drawn as 4 hundreds blocks, 6 tens blocks, and 2 ones blocks.

8　Answer: B

Explanation:　Determine the age of the school by subtracting 10 from 150.

9　Answer: C

Explanation:　Determine the age of the table by subtracting 100 from 275.

10　Answer: 305

Explanation:　Represent the situation with the number sentence $325 - 10 - 10 = 305$.

11　Answer: 125

Explanation:　Represent the situation by the equation $225 - 100 = 125$.

12　Answer: 750

Explanation:　First find the number of calories in Paul's snack $430 + 220 = 650$. Then add 100 to find the number of calories in Randy's snack.

13　Answer: 100

Explanation:　Tao eats $100 + 100 + 100 = 300$ calories. Mari eats $100 + 100 + 100 + 100 = 400$ calories. $400 - 300$ is 100 calories.

14　Answer: 200

Explanation:　Add 10 to 190 to get 200.

15　Answer: C

Explanation:　$100 + 45$ people like chicken nuggets and pizza. 45 in expanded form is $40 + 5$. So $100 + 40 + 5$ is the total number of people that like pizza and chicken nuggets. 86 people like tacos. 86 in expanded form by $80 + 6$. Subtract the number of people who like tacos from the total number of people who like pizza and chicken nuggets.

16　Answer: 20 and 23

Explanation:　The pattern between each number is add 3. The values in the list increase by 3 each time.

17　Answer: 110 and 135

Explanation:　The pattern between each number is add 25. The values in the list increase by 25 each time.

18　Answer: 74; Explanation will vary

Explanation:　Answers must include a reasonable explanation such as, "Joe and Tao swim a total of 74 minutes. I know because Tao swims for 32 minutes and Joe swims for 10 more minutes, so $32 + 10 = 42$. Joe swims for 42 minutes. To find how long they swam together, add $32 + 42 = 74$.

19　Answer: Nap; Explanation will vary

Explanation:　Answers must include a reasonable explanation such as, "The dog will spend more time taking a nap than digging holes. I know because the dog spends 29 minutes digging holes. I found this by subtracting 16 and 30 from 75. The dog takes a nap for 30 minutes which is greater than 29 the amount of time the dog digs holes."

20　Answer: 500

Explanation:　The pattern increases by 75, and the next number in the pattern is 500. Answers must include a reasonable explanation such as, " 500 is the next number. I know because the pattern is adding 75 and $75 + 425 = 500$."

NUMBER AND OPERATIONS IN BASE TEN CHAPTER REVIEW

1　Answer: 138

Explanation:　Answers must include a reasonable explanation such as, "The final score was 138. I know because the number 138 has a one in the hundreds place, a 3 in the tens place, and an 8 in the ones place."

ANSWERS and EXPLANATIONS

2 Answer: B

Explanation: 700 is 7 hundreds.

3 Answer: 0

Explanation: 340 has a zero in the ones place.

4 Answer: C

Explanation: The orchard sold 450 + 370 = 820 bags of apples altogether.

5 Answer: B

Explanation: When counting by fives 845 comes after 840.

6 Answer: 710

Explanation: The count sequence is by 5's. 710 is immediately before 715 when counting by fives.

7 Answer: A

Explanation: Nine hundred eighty- nine is the numeral 989.

8 Answer: D

Explanation: Eight hundred forty-four written in expanded form is 800 + 40 + 4.

9 Answer: Disagree; Explanation will vary

Explanation: Answers must include a reasonable explanation such as, "Max is correct that 784 is less than 857, however, he used the wrong sign. He should have written 784 < 857.

10 Answer: A

Explanation: 645 is less than 654, so 645 < 654. The inequality symbol always points to the smaller side.

11 Answer: 4

Explanation: The tallest student is 50 inches tall, and the shortest student is 46 inches tall. The difference is: 50 – 46 = 4.

12 Answer: 61

Explanation: Answers must include a reasonable explanation such as, "The truck and car weigh 61 grams together. I know because two cars weigh 58 grams and 29 +

29 equals 58. So one car weighs 29 grams. The truck weighs 32 grams. Together the car and the truck weigh 61 grams because 29 + 32 = 61.

13 Answer: mittens, hat, shirt, and glasses

Explanation: The student will select a combination of items which have a total cost that is less than $65.

14 Answer: 72

Explanation: The image shows 12 stamps on each page. There will be a total of 6 pages in the book: 12 added 6 times is 72.

15 Answer: C

Explanation: Subtract the "In" number from the "Out" number. Adding 188 is the rule.

16 Answer: 90

Explanation: First subtract 150 and 90. 150 -90 =60. Then add 30 to 60. 60 + 30 = 90.

17 Answer: 76

Explanation: The student is adding 2 groups of 10 (or 20) books to the original amount. 56 + 10 + 10 = 76.

18 Answer: 107

Explanation: Roni's brother has 100 + 17 = 117 rocks. Roni's sister has 117 – 10 = 107 rocks.

19 Answer: Disagree; Explanation will vary

Explanation: Lau's decomposition strategy is correct. The tens and ones can be added separately. Lau makes a computational error when adding 80 + 90 + 60 + 100. The sum is 330, not 230.

20 Answer: 17 and 25; Explanation will vary

Explanation: All three boxes together weigh 94 pounds. The total 94 minus the weight of the large box 52 equals 42. 42 – 8 = 34 and 17 + 17 equals 34. Therefore the smallest box weighs 17 pounds and the middle box weighs 17 + 8 or 25 pounds.

prepaze

ANSWERS and EXPLANATIONS

NUMBER AND OPERATIONS IN BASE TEN EXTRA PRACTICE

1 Answer: Answers will vary

Explanation: Answers must include a reasonable explanation such as, "I would tell Edwin to draw 3 hundreds blocks, 7 tens blocks, and 1 ones block."

2 Answer: 90

Explanation: Answers must include a reasonable explanation such as, "James has 204 cubes so far. He needs to add 9 tens cubes to have 294."

3 Answer: C

Explanation: 94 is 0 hundreds, 9 tens, and 4 ones.

4 Answer: 495

Explanation: 495 is 5 before 500.

5 Answer: 900

Explanation: 9 hundreds is 900.

6 Answer: D

Explanation: Jessica will say 500, 600, and then 700 or she will say 700 is three hundreds after 400.

7 Answer: Disagree

Explanation: Answers must include a reasonable explanation such as, "I disagree. My partner forgot to split the tens and ones. It should be 600 + 20 + 5."

8 Answer: B

Explanation: 800 + 0 + 6 is the same as 806.

9 Answer: B

Explanation: The figure shows 700 + 50 + 4; 746 is less than the 754 cubes that are shown.

10 Answer: =

Explanation: 640 is equal to 640.

11 Answer: A

Explanation: Larry's dog weighs 60 pounds (75 − 15) and Cleo's dog weighs 60 + 10 = 70 pounds.

12 Answer: A

Explanation: The correct first step is to subtract 24 from 37 because the problem only involves adding and subtracting. Do the arithmetic from the left to the right.

13 Answer: A

Explanation: Tim has 20 + 15 = 35 candies. Find the total number of candies by adding 15 + 22 + 14 + 35 = 86.

14 Answer: Barry – 18, Aja – 14, Chuck – 28, Eric – 10; together – 70

Explanation: The student should use order of numbers sense and reasoning to determine the number of toys each person has. Then add the number of toys.

15 Answer: D

Explanation: Three hops of 240 is 240 + 240 + 240 = 720. 920 − 720 is 200.

16 Answer: Height: 403 Width: 451

Explanation: The new height is 278 + 125 = 403. The new width is 343 + 108 = 451.

17 Answer: 175

Explanation: 375 minus 100 minus 100 equals 175.

18 Answer: 480

Explanation: Luke has 500 dollars. 500 minus 10 minus 10 equals 480.

19 Answer: Agree; Explanation will vary

Explanation: Answers must include a reasonable explanation such as, " Rina is correct because she started at 719 then subtracted 100 and 5 which is 105. Next she added 100 and 100 which is 200."

20 Answer: Explanation will vary
Explanation: The student may identify the similarities and differences in structure / composition, the number of terms, the types of numbers and strategies used to add these numbers. For example, 79 is close to 80, so by rounding it to 80, the expression could be thought of as (80 + 100) -1 or 180 -1. With the expression 400-128, 128 is also close to 130. Rounding 128 to 130 makes the expression (400 − 130) + 2.

MEASUREMENT AND DATA
UNIT 1: MEASURING LENGTH

1 Answer: B
Explanation: The length of the block is 11 centimeters.

2 Answer: C
Explanation: The length of the trumpet is 12 centimeters.

3 Answer: 11
Explanation: The length of the black bar is 16 cm, and the length of the white bar is 5 cm. 16 minus 5 equals 11.

4 Answer: C
Explanation: The expression, 80-20, is equivalent to the length of the bar, which is 60 units.

5 Answer: D
Explanation: The sword is 5 centimeters long.

6 Answer: C
Explanation: The crown is 2 centimeters long.

7 Answer: B
Explanation: The dog is 5 centimeters long.

8 Answer: B
Explanation: The length of the line is equal to two dice or one paper clip.

9 Answer: D
Explanation: The length of the line is equal to 3 dice or 3 buttons.

10 Answer: C
Explanation: The length of the line is equal to 6 buttons or 3 paper clips.

11 Answer: C
Explanation: 6 yards is the best estimate. The other choices are too short or too long.

12 Answer: A
Explanation: 9 feet is the best estimate. The other choices are short or too long.

13 Answer: B
Explanation: 30 yards is the best estimate. The other choices are too short or too long.

14 Answer: A
Explanation: Add 1 foot to the other height, 3 feet, to determine the height of the soccer poster.

15 Answer: C
Explanation: 7 minus 1 equals 6.

16 Answer: B
Explanation: The blue house is 26 feet tall, the red house is 23 feet tall, and the beige house is 25 feet tall. Therefore, the blue house is the tallest.

17 Answer: A
Explanation: 36 minus 28 equals 8.

18 Answer: A
Explanation: 78 minus 13 equals 65.

19 Answer: B
Explanation: The building is 60 − 12 = 48 meters taller not 34.

20 Answer: 700
Explanation: 1000 minus 300 equals 700.

prepaze

MEASUREMENT AND DATA
UNIT 2: TIME AND MONEY

1 Answer: C
Explanation: Music class starts at 2:05. If the class lasts 40 minutes, it ends at 2:45 pm.

2 Answer: B
Explanation: Bill arrived at his Grandma's house at 11:05. He left his house 15 minutes before that, which means he left his house at 10:50. The hour hand will be past the 10, and the minute hand will be on the 10.

3 Answer: D
Explanation: 30 minutes before 11:10 is 10:40.

4 Answer: B
Explanation: It takes her 25 minutes to travel home. 25 minutes after 3:45 is 4:10.

5 Answer: 10:25
Explanation: Baking two pans of muffins takes 70 minutes. Seventy minutes past 9:15 am is 10:25 am. The little hand should be past 10 and the big hand should be on the 5.

6 Answer: 3:05
Explanation: When drawing the analog clock, the hour hand should be on the 3, and minute hand should be on the 1. The digital clock should say 3:05.

7 Answer: 6:20
Explanation: 1 hour and 20 minutes have passed.

8 Answer: B
Explanation: The clock shows 1:10. One hour later is 2:10.

9 Answer: 6:30
Explanation: "Half past six" is 6:30. The large hand on the clock is on the 6 and the small hand is halfway between 6 and 7.

10 Answer: 35
Explanation: Answers must include a reasonable explanation such as, "35 minutes have passed. I know because I counted by fives starting at the 12 and ending at the 7.

11 Answer: B
Explanation: The coins shown have a value of $0.22. Adding $0.58 means there is $0.80.

12 Answer: C
Explanation: Quarters are 25. Count the 2 quarters first (25, 50. Dimes are 10. Count the 6 dimes (60, 70, 80, 90, 100, 110). Nickels are 5. Count the 3 nickels (115, 120, 125), Umar has 125 cents.

13 Answer: $2.45
Explanation: Joey has $1.10 and Kari has $1.35. Their combined total is $2.45.

14 Answer: D
Explanation: Miles has $3.45. The chicken nuggets and fries will cost $3.15. The other choices cost less or too much.

15 Answer: Jin
Explanation: Hyun has $6.00. Jin has $6.25.

16 Answer: C
Explanation: The value of coins and bills she put in the account is $4.03. Subtract $4.03 from $10.70 which equals $6.67.

17 Answer: B
Explanation: Diego has 65 cents. Hugo has 75 cents. Jada has $1. Altogether they have $2.40.

18 Answer: $0.86
Explanation: 4 quarters, 3 dimes, 8 nickels, and 2 pennies is worth $1.72. Half of $1.72 is $0.86.

19 Answer: Dara
Explanation: Answers must include a reasonable explanation such as, " Dara has more money because she has $2.40 and Chin has $1.51. $2.40 is greater than $1.51 therefore, Dara has more money than Chin."

20   Answer: Explanations will vary, $0.65 left

Explanation: Answers must include a reasonable explanation such as, "Fredo can exchange his one-dollar bill for 20 nickels. This would allow him to give his friend 7 nickels, or $0.35, and he would have $0.65 remaining."
or any combination of nickels, dimes, and quarters that will allow $0.35 in change.

MEASUREMENT AND DATA
UNIT 3: LEARN TO READ DATA

1   Answer: C

Explanation: The height of the kitten's at 2 months is 12 inches, at 4 months it is 18 inches, and in 5 months it is 21 inches tall.

2   Answer: A

Explanation: The objects are small because they are between 3 and 7 inches. Therefore, only pencils would be correct.

3   Answer: B

Explanation: The length of the items are 6 inches, 5 inches, 5 inches and 2 inches. Therefore, there should be one x at 2, two x's at 5 and one x at 6.

4   Answer: B

Explanation: The lengths of the pencils are: 5 inches, 4 inches, 8 inches, and 7 inches.

5   Answer: A

Explanation: The objects are small because they are between 6 and 9 inches. Therefore, only scissors is correct. The other objects are too large.

6   Answer: 10

Explanation: The longest snake is 24 inches and the shortest snake is 14 inches. 24 minus 14 equals 10.

7   Answer: 7

Explanation: Line plot should show 2 marks at 5 inches, 3 marks at 7 inches, 5 marks at 8 inches, and 2 marks at 9 inches. There are 7 pencils that are 8 inches or longer.

8   Answer: 2, 8, 10, 4, 5

Explanation: Use the rulers to measure the length of each leaf, then record the data in the table.

9   Answer: B

Explanation: The 4 data points for the line plot are 18, 19, 20, and 20. The scale of the line plot should be correctly labeled with the weight.

10   Answer: 14

Explanation: The line plot should look like

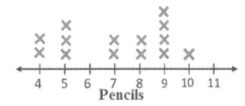

Add the numbers in the columns for the number of pencils. 2 + 3 + 2 + 2 + 4 + 1 = 14

11   Answer: 51

Explanation: Each icon is 2 cones and each half icon is 1 cone. Count the icons, multiply the full icons by 2, and the half icons by 1. Add the results: 16, 10, 15 and 10.

12   Answer: D

Explanation: There are 9 apples collected Wednesday. There are 9 + 5 or 14 apples collected Friday.

13   Answer: Spiders, Explanations will vary

Explanation: Answers must include a reasonable explanation such as, "Kyle saw the least spiders because the graph shows he saw only 3 of them where he saw more than 3 of each of the other insects.

14   Answer: D

Explanation: 9 people like baseball and 4 people like basketball. 9 minus 4 equals 5.

15 Answer: 92

Explanation: Each icon represents 8 miles. The graph contains 11.5 icons. Multiply 8 by 11.5 which results in 92 miles.

16 Answer: 7 robins, 9 bluebirds, 2 sparrows, 5 cardinals.

Explanation: The scale on the bar graph is increasing by 1s. The heights in the bar graph are 5 for cardinals, 2 for sparrows, 7 for robins, and 9 for bluebirds.

17 Answer: 10

Explanation: The owl travels a total of 84 miles over four days. On Monday the owl traveled 12 miles, Tuesday 18 miles, Thursday 24 miles. 12 + 18 + 24 equals 54. 84 minus 54 equals 30. Each owl represents 3 miles so add 10 threes for 30 more miles.

18 Answer: No

Explanation: Answers must include a reasonable explanation such as, "I disagree with Tommy because each cake symbol equals 3 people. Therefore, 12 people like vanilla cake and 3 people like chocolate cake. 12 minus 3 is 9. So 9 more people like vanilla than chocolate.

19 Answer: C

Explanation: 11 people like red and yellow. 8 people like pink and purple. 11 − 8 = 3.

20 Answer: 2.5 trees for birch and 2 trees for pecan

Explanation: Each symbol represents 4 trees. The birch trees would be represented with 2½ symbols. The pecan tree would be represented with 2 symbols.

MEASUREMENT AND DATA CHAPTER REVIEW

---

1 Answer: B

Explanation: Soccer is currently the favorite sport with 13 people which means that football would need to have 14 people. 14 minus 4 equals 10.

2 Answer: 9

Explanation: Each icon represents 2 cones. There were 20 cones sold on Wednesday and 11 cones sold on Tuesday. 20 minus 11 equals 9.

3 Answer: A

Explanation: The height of the dogs is being measured with non-standard units (blocks); each unit represents 2 inches. Count the blocks and multiply by 2.

4 Answer: 6

Explanation: 9 people jumped more than 15 feet and 3 people jumped 15 feet. 9 minus 3 equals 6.

5 Answer: A

Explanation: One quarter equals $0.25, one dime equals $0.10, and two nickels equal $0.10. $0.25 plus $0.10 plus $0.10 equals $0.45.

6 Answer: D

Explanation: Her sister gives her $0.54, so her brother gives her $1.08. $1.75 plus $0.54 plus $1.08 equals $3.37.

7 Answer: B

Explanation: The hour hand is past the 12, which means it is right after 12 o'clock. The minute hand points to the 7 and this represents 35 minutes. The time is 12:35.

8 Answer: Agree; Explanation will vary

Explanation: Answers must include a reasonable explanation such as, "I agree with Bomi because the little hand is between 11 and 12 which mean it is after the 11 o'clock hour. Then 5 added 9 times is 45, so the minutes are 45. Bomi is correct it is 11:45."

9 Answer: B

Explanation: The green pencil is 7 units; 2 to 9 is 7 spaces. The yellow pencil is 5 units; 9 to 14 is 5 spaces. 7 plus 5 equals 12.

10   Answer: Answers will vary
Explanation: The student should draw a line segment which has a length of 7 units. The line segment can start at any point between 15 and 35 to fit on this number line.

11   Answer: B
Explanation: Subtract: 70 minus 58 equals 12 inches.

12   Answer: A
Explanation: Total length of the pencils is 17 + 12 + 9 = 38 cm.

13   Answer: D
Explanation: Add 2 inches to 8 inches: 10 inches.

14   Answer: B
Explanation: Caroline's bedroom is 14 ft long and Jameson's is 16 ft long. 16 is larger than 14.

15   Answer: B
Explanation: 2 inches is the best estimate. The other choices are too long.

16   Answer: D
Explanation: 1 foot is the best estimate. The other choices are too small or too large.

17   Answer: B
Explanation: The screw is 5 dice long, but only 5 buttons long, not 6.

18   Answer: 2 , 4
Explanation: The shoe is 2 paper clips long and 4 buttons long.

19   Answer: B
Explanation: The car is 5 ½ units long.

20   Answer: C
Explanation: Of the choices, only a clock measures time. It measures hours, minutes, and seconds.

MEASUREMENT AND DATA EXTRA PRACTICE

1   Answer: C
Explanation: A stopwatch can be used to measure the time.

2   Answer: C
Explanation: A tape measure can be used to measure length or height.

3   Answer: 2 ; 4
Explanation: The lizard is 2 paper clips or 4 dice long.

4   Answer: 3 ; 6
Explanation: The length of the ice cream is 3 paper clips or 6 dice.

5   Answer: C
Explanation: 14 yards is the best estimate for the length of the bus. The other choices are too short or too long.

6   Answer: A
Explanation: 2 meters is about 6 feet. This is a good height estimate.

7   Answer: 5 feet 5 inches
Explanation: 3 inches + 2 inches = 5 inches.

8   Answer: 5 meters
Explanation: 3 meters + 2 meters = 5 meters.

9   Answer: 24
Explanation: 56 – 32 = 24 centimeters.

10   Answer: 34
Explanation: 19 + 15 = 34 centimeters.

11   Answer: C
Explanation: One cylinder is 8 units wide. Two cylinders is 8 + 8 units or 16 units wide.

12   Answer: C
Explanation: The piece of string shown is 13 inches. 13 + 13 + 13 + 13 represents the length of the entire string.

ANSWERS and EXPLANATIONS

13   Answer: A
Explanation: Joni will arrive home at 3:00. The hour hand will be on the 3 and the minute hand will be on the 12.

14   Answer: 3:30; Explanations will vary
Explanation: Answers must include a reasonable explanation such as, "Kari's cat finished eating at 3:30 pm. I know because the dog finished eating at 3:00 pm. Adding 45 minutes to 2:15 is 3:00 pm. The cat takes 30 minutes to finish its food, which means it finishes eating at 3:30 pm."

15   Answer: B
Explanation: Dev has $1.98, Evie has $1.50, Josie has $1.30, and Liam has $0.50.

16   Answer: $4.20
Explanation: Ken puts $1.05 in his bank on Monday. Each following day, he adds $1.05, which means after 4 days, he will have a total of $4.20.

17   Answer: 2 and 3
Explanation: Statement 1 is false because only 3 students live less than 3 miles from school. Statement 2 is true because the line plot shows 7 students live more than 3 miles from school. Statement 3 is true because 1 more student lives 3 miles from school than those who live 2 miles from school.

18   Answer: Explanations will vary
Explanation: Answers must include a reasonable explanation such as, "The data in the bar graph shows the temperature in each city is as follows: 50, 25, 70, 30, 45, 40. Each of these values should be represented as a data point on the line graph. The line graph should have 6 data points. The frequency of each piece of data is denoted with an X on a line plot, where on the bar graph, it is a bar with a length corresponding to the number of data points."

19   Answer: B
Explanation: 8 people voted for red (the favorite color), and 2 people voted for yellow (the least favorite color). 8 minus 2 equals 6.

20   Answer: Answers will vary
Explanation: Answers must include a reasonable explanation such as, "I would make The categories: lions, monkeys, tigers, and giraffes. The bar graph would need to show 4 lions, 3 monkeys, 3 tigers, and 2 giraffe."

GEOMETRY
UNIT 1: UNDERSTANDING SHAPES

1   Answer: 3
Explanation: There is 3 squares in both circles. They are in the center of the Venn diagram.

2   Answer: 6
Explanation: There are 3 squares in both circles and 3 squares in the right circle, for a total of 6 squares.

3   Answer: 3
Explanation: There are 3 squares that are in the right circle, but not in the left circle. The squares that are in the left circle appear in the center portion of the Venn diagram.

4   Answer: 3
Explanation: There are 2 triangles and 1 circle, for a total of 3 shapes in the left circle.

5   Answer: 5
Explanation: There are 3 circles in the center and 2 circles in the right circle, for a total of 5 circles.

6   Answer: 3
Explanation: The shape is a triangle. Therefore, it has 3 sides.

7   Answer: Cube
Explanation: The cube has 6 square faces.

8   Answer: Rectangular prism
Explanation: The rectangular prism has 4 faces which are the same length and width as the 2-D rectangle.

9 Answer: Cylinder

Explanation: The cylinder has circular bases.

10 Answer: 3

Explanation: The vertices are where each edge (side) meets. A triangle has 3 vertices.

11 Answer: 5

Explanation: A pentagon has 5 sides.

12 Answer: 5

Explanation: A pentagon has 5 vertices.

13 Answer: 4

Explanation: A rhombus has 4 sides.

14 Answer: 4

Explanation: A rhombus has 4 vertices.

15 Answer: 6

Explanation: A hexagon has 6 sides.

16 Answer: 6

Explanation: A hexagon has 6 vertices.

17 Answer: Diamond to the kite.

Explanation: A line is drawn from the diamond to the kite because the kite is diamond shaped.

18 Answer: Circle to the clock.

Explanation: A line is drawn from the circle to the clock because the clock is a circle.

19 Answer: Triangle to the emergency sign.

Explanation: A line is drawn from the triangle to the emergency sign because the sign is in the shape of a triangle.

20 Answer: Rectangle to the door.

Explanation: A line is drawn from the rectangle to the door because the door is in the shape of rectangle.

GEOMETRY
UNIT 2: PARTITION SHAPES INTO ROWS AND COLUMNS

1 Answer: B

Explanation: An array with 2 rows and 3 columns has 2+2+2 squares or 6 total squares.

2 Answer: C

Explanation: The pan of brownies has 3 horizontal rows and 4 vertical columns.

3 Answer: D

Explanation: To have 8 equal size pieces, the chocolate bar could have 4 rows and 2 columns.

4 Answer: B

Explanation: The array shows 3 horizontal rows with 5 vertical columns.

5 Answer: B

Explanation: The array of 9 squares can be described as 3 rows of 3 squares.

6 Answer: A

Explanation: The array has 3 horizontal rows and 4 vertical columns.

7 Answer:

Explanation: The array has 3 horizontal rows and 6 vertical columns and a total of 18 squares.

8 Answer: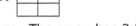

Explanation: The array has 2 horizontal rows with 2 squares in each row.

9 Answer: Neither

Explanation: They both have 12 squares.

10 Answer: Ivan

Explanation: Ivan's array has 16 squares and Gino's has 10 squares.

prepaze

11 Answer:

Explanation: The array has 5 horizontal rows and 4 vertical columns and a total of 20 squares.

12 Answer:

Explanation: The array has 6 horizontal rows and 3 vertical columns and a total of 18 squares.

13 Answer:

Explanation: The array should have 2 rows with 4 columns, or 4 rows with 2 columns.

14 Answer:

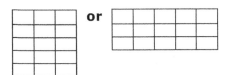

Explanation: The array should have 3 rows with 5 columns or 5 rows with 3 columns.

15 Answer:

Explanation: The array should have a total of 16 squares, in 4 equal rows and columns.

16 Answer:

Explanation: The array should have a total of 3 squares, in one column.

17 Answer: A
Explanation: The figure has 3 squares.

18 Answer: A
Explanation: The figure has 4 squares.

19 Answer: C
Explanation: The figure has 12 squares.

20 Answer: C
Explanation: The figure has 6 squares.

GEOMETRY
UNIT 3: PARTITION SHAPES INTO EQUAL PARTS

1 Answer: B
Explanation: Both parts are the same size and shape.

2 Answer: A and C
Explanation: The parts in each shape are the same size.

3 Answer: A, B, and D
Explanation: The parts in each shape are the same size.

4 Answer: A and B
Explanation: The parts in each shape are the same size.

5 Answer: A
Explanation: The 4 parts are the same size.

6 Answer: A
Explanation: The cookie is divided into 2 equal parts.

7 Answer: B
Explanation: The cookie is divided into 4 equal parts.

8 Answer: A
Explanation: The cupcake is divided into 2 equal parts.

9 Answer: B
Explanation: The pizza is divided into 4 equal parts.

ANSWERS and EXPLANATIONS

10   Answer: B

Explanation: The watermelon is divided into 2 equal parts.

11   Answer: A

Explanation: The rectangle is divided into 2 equal parts.

12   Answer: C

Explanation: The rhombus is divided into 4 equal parts. One part is green.

13   Answer: B

Explanation: The circle is divided into 3 equal sized parts. One part is shaded.

14   Answer: A

Explanation: The rectangle is divided into 2 equal parts.

15   Answer: C

Explanation: The rectangle is divided into 4 equal parts.

16   Answer: D

Explanation: The circle is divided into 4 equal sized parts.

17   Answer: A

Explanation: The rectangle is divided into 2 equal parts and 2 of the parts are shaded.

18   Answer: D

Explanation: The circle is divided into 3 equal parts and 3 of the parts are shaded.

19   Answer: A

Explanation: The square is divided into 4 equal parts. One part is shaded.

20   Answer: D

Explanation: The rectangle is divided into 4 equal parts. Four parts are shaded.

GEOMETRY CHAPTER REVIEW

1   Answer: B

Explanation: The square 4 sides and 4 vertices, which is more than 3 angles and less than 6 sides.

2   Answer: B

Explanation: The array has 3 rows of 4 squares and a total of 12 parts.

3   Answer: C

Explanation: The circle is divided into two equal parts.

4   Answer: A

Explanation: A triangle has fewer sides than a rectangle.

5   Answer: A

Explanation: The array is a 5 by 3 therefore, has 15 pieces.

6   Answer: C

Explanation: The circle is cut into 3 equal parts and 3 of the parts are shaded.

7   Answer: C

Explanation: There are 6 faces on a cube, top, bottom, right side, left side, front, and back.

8   Answer: A

Explanation: Four rows of 2 creates 8 square parts.

9   Answer: C

Explanation: Dividing a whole into 4 equal parts creates fourths.

10   Answer: A

Explanation: A quadrilateral has 4 sides and a pentagon has 5 sides.

11   Answer: B

Explanation: An array of 6 rows and 1 column will create 6 pieces.

prepaze

12 Answer:

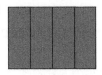

Explanation: The rectangle should be divided into 4 equal sized parts with all parts shaded.

13 Answer: Yes; Explanation will vary

Explanation: Answers must include a reasonable explanation such as, "Kwan is correct because each of these prisms is a rectangular prisms. All rectangular prisms all have 6 faces: top, bottom, right side, left side, front, and back.

14 Answer: D

Explanation: Sharing the chocolate bar into 12 equal parts can be done by Dividing it into 4 rows and 3 columns.

15 Answer:

Explanation: The circle should be partitioned into 3 equal sized parts with all parts shaded.

16 Answer: Answers will vary

Explanation: Solutions include a rectangle, a kite, a trapezoid, or any quadrilateral as long as two sides have different lengths.

17 Answer: 3 + 3 + 3 = 9

Explanation: The array shows 3 rows of 3 squares which has 9 squares.

18 Answer: One-third

Explanation: Solutions include 1/3, one-third, or a third.

19 Answer: No

Explanation: A pentagon has 5 vertices. A hexagon has 6 vertices.

20 Answer: 3 + 3 + 3 + 3 + 3 = 15

Explanation: The array shows 5 rows of 3 squares and has 15 squares.

GEOMETRY
EXTRA PRACTICE

1 Answer: C

Explanation: A cube has 12 edges and 6 faces: top, bottom, right side, left side, front, and back.

2 Answer: A

Explanation: 2 rows with 7 columns will create 14 squares.

3 Answer: C

Explanation: The partitions divide the rectangle into thirds. The shaded part is half of a third (or one-sixth).

4 Answer: D

Explanation: The triangle requires 3 toothpicks. The hexagon requires 6 toothpicks. 3 + 6 = 9

5 Answer: B

Explanation: Both arrays have a total of 12 squares.

6 Answer: A

Explanation: There are 3 equal sized parts and one part is shaded.

7 Answer: B

Explanation: A rectangle has 4 interior angles and 4 corners.

8 Answer: D

Explanation: There is 1 row of 9 squares in the array.

9 Answer: B

Explanation: The rectangle has 4 equal sized parts.

10 Answer: Answers will vary

Explanation: Answers must include a reasonable explanation such as, "The shapes have 4 sides. The shapes all have 4 vertices. The shapes are all quadrilaterals."

11 Answer: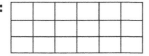

Explanation: The array should have 1 horizontal row of 4 squares.

12 Answer: Neither
Explanation: Three thirds and two halves are both equivalent to one whole.

13 Answer: A and B
Explanation: Pentagons are polygons with 5 sides.

14 Answer: Eric
Explanation: Eric's array has 15 squares and Hal's array has 12 squares.

15 Answer:

Explanation: The circle should have 2 equal parts. Both parts should be shaded.

16 Answer: B
Explanation: Shape B has 12 sides.

17 Answer:

Explanation: The array should have 3 horizontal rows and 6 columns. There should be 18 squares total.

18 Answer: Yes
Explanation: Answers must include a reasonable explanation such as, "Both rectangles have 4 parts. All four parts are shaded."

19 Answer: B
Explanation: The hexagon has 6 sides.

20 Answer:

Explanation: The array should have 2 horizontal rows with 5 squares in each row.

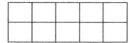

COMPREHENSIVE ASSESSMENTS
ASSESSMENT 1

1 Answer: 6
Explanation: 53 plus 31 equals 84. The farmer has picked 84 vegetables. $90 - 84 = 6$.

2 Answer: 5
Explanation: There are 5 rows, so she should add the number 2 five times.

3 Answer: 705, Explanation will vary
Explanation: Answers must include a reasonable explanation such as, "You are thinking of 705. I know because 705 have 7 hundreds, 0 tens and 5 ones, or 700 + 0 + 5."

4 Answer: 540, Explanation will vary
Explanation: Answers must include a reasonable explanation such as, "Jeremy has 540 pennies. If I count by 10s 540 comes right after 530 or if I add 10 to 530, the answer is 540."

5 Answer: A
Explanation: The array shows 4 rows of 4 squares.

6 Answer: B
Explanation: The whole shows 2 equal parts.

7 Answer: 837; Explanation will vary
Explanation: Answers must include a reasonable explanation such as, "800 + 30 + 7 is the expanded form of 837. I know because 800 is 8 hundreds, 30 is 3 tens and 7 is seven ones. 8 hundreds, 3 tens and 7 ones is 837."

8 Answer: Disagree; Explanation will vary

Explanation: Answers must include a reasonable explanation such as, "I disagree. To compare these numbers, you should look at the hundreds place; 6 hundreds are more than 7 hundreds, so you don't need to look at the tens."

9 Answer: A

Explanation: $9 + 6 = 15$

10 Answer: No; Explanation will vary

Explanation: Answers must include a reasonable explanation such as, "I disagree with Sky. There are more than 14 flowers drawn. An appropriate equation could be $(7+7) +1 = 15$

11 Answer: 468, 97, 595, 894

Explanation: Subtract 106 from each input "In" value to get each "Out" value.

12 Answer: $25

Explanation: $\$225 - \$100 - \$100 = \$25$

13 Answer: A

Explanation: The associative property; combining numbers which can be used to make a ten is being applied to $71 + 39$. Then 55 is added to it.

14 Answer: A

Explanation: The line is 2 cm long.

15 Answer: A

Explanation: The line is the length of 1 die and 1 button.

16 Answer: A

Explanation: 17 yards is the best estimate. The other choices are too short or too long.

17 Answer: 81

Explanation: The scale reads $75 + 6 = 81$

18 Answer: Yes; Explanation will vary

Explanation: Answers must include a reasonable explanation such as, "I agree

with Zara because she regroup into 10s. Her number line shows $50 + 10 + 10 + 10 = 70$ and $6 + 1 + 3 = 10$, so $70 + 10 = 80$. Also, $56 + 11 + 13 = 80$. Therefore, Zara is correct."

19 Answer: B

Explanation: Andrea's sailboat is 16 feet long.

20 Answer: No, Explanations will vary

Explanation: Answers must include a reasonable explanation such as, "I disagree, Taru had 9 hundreds, 9 tens, and 4 ones. Therefore, her final answer should have been 994."

21 Answer: C

Explanation: The distance from 3 to 11 is 8 units.

22 Answer: $40 + 40 - 60 + 40$; Explanations will vary

Explanation: Answers must include a reasonable explanation such as, " Each hash mark is 20. The expression starts at 40 then 40 is added to 40 to get to 80. Next, 60 is subtracted from 80 to equal 20. Finally 40 is added to 20 to equal 60."

23 Answer: Melissa

Explanation: Answers must include a reasonable explanation such as, " Melissa spent more money because 25 plus 31 plus 29 equals 85 and 85 dollars is more than 75 dollars.

24 Answer: B

Explanation: The hour hand is past the 10, and the minute hand is on the 10 which is equal to 50 minutes. The time is 10:50.

25 Answer: B

Explanation: $42 + 18 = 60$ cm

26 Answer: B

Explanation: From the graph, find the number of animals by adding 10, 14, 7, and 15. $(10 + 14 + 7 + 15 = 46)$. Add 5 gives 51.

27 Answer: A
Explanation: A triangle is a two-dimensional shape. The other shapes are three-dimensional shapes.

28 Answer: Yes; Explanation will vary
Explanation: Answers must include a reasonable explanation such as, "I agree because Lee used regrouping to get groups of 10s. 20 + 50 + 30 − 6 = 94 And 17 + 48 + 29 = 94.

29 Answer: 7
Explanation: 7 plus 7 equals 14.

30 Answer: 15
Explanation: 3 rows and 5 columns is 15.

31 Answer: C
Explanation: There are 6 hundreds in 673.

32 Answer: B
Explanation: When counting by 5's, 990 comes after 985.

33 Answer: Six hundred ninety-three
Explanation: Six hundred ninety-three is the word form of 693.

34 Answer: A
Explanation: Pari has $1.75. Fries and milk cost $1.70.

35 Answer: 981 > 822
Explanation: Nine hundred eighty-one is greater than eight hundred twenty-two can be written as 981 > 822. The symbol always points to the smaller side.

36 Answer: D
Explanation: The measurements of the objects are: 2 in, 6 in, 8 in, 8 in.

37 Answer: 175 + 86 + ? = 400; 139
Explanation: 400 − 175 − 86 = 139

38 Answer: A
Explanation: 80 + 70 + 90 = 240 kilograms.

39 Answer: 7:45
Explanation: Subtract 35 minutes from 8:30 pm = 7:55. Subtract 10 minutes from 7:55 pm for 7:45 pm.

40 Answer: B and C
Explanation: Answers must include a reasonable explanation such as, "B and C are correct because the distance from 4 to 20 is 16. the distance from 0 to 16 is 16. So both B and C are 16 line segments"

41 Answer: $2.15
Explanation: Raley has $10.00, she gives Lara $5.70 so she has $4.30 left. Then she give half of that to her sister, leaving her with $2.15.

42 Answer: 490
Explanation: Point A represents 600. So, 600 − 100 − 10 = 490.

43 Answer: D
Explanation: Using compatible numbers: 79 and 11 can be grouped together because they make a tens; 25 and 25 can be grouped because it represents a double.

44 Answer: 9; Explanation will vary
Explanation: Answers must include a reasonable explanation such as, "There are 9 data points, which means the line plot shows there are 9 days."

45 Answer: 10
Explanation: The bar graph shows there are 4 rose bushes planted. Six more than 4 is 10.

COMPREHENSIVE ASSESSMENTS
ASSESSMENT 2

1 Answer: 96
Explanation: 32 plus 4 plus 60 (6 x 10) equals 96.

2 Answer: 25
Explanation: 5 groups of 5 is 25.

prepaze

3 Answer: 250

Explanation: 250 is 2 hundreds and 5 tens, 200 + 50.

4 Answer: D

Explanation: Sophia has 250 + 220 = 470 marbles.

5 Answer: D

Explanation: The array shows 1 row of 15 columns.

6 Answer: C

Explanation: The basketball is divided into 4 equal parts. Each part is a fourth of the ball.

7 Answer: 900 + 30 + 1

Explanation: 900 + 30 + 1 is the expanded form of 931.

8 Answer: B

Explanation: 870 is greater than 807

9 Answer: D

Explanation: $17 - 14 = 3$

10 Answer: Yes; Explanations will vary

Explanation: Answers must include a reasonable explanation such as, "When you count the cupcakes by 2s you have 9 sets of 2s. The doubles fact 9 plus 9 to represents the picture. There are 18 cupcakes in the picture."

11 Answer: Answers will vary

Explanation: Strategies should involve or describe regrouping for addition. Standard algorithm to add ones (4 + 5 = 9), tens (70 + 30 = 100, or regrouping to show there are 0 tens), and hundreds (400 + 100 + 100 = 600). The sum is 609.

12 Answer: 502 and 900

Explanation: Answers must include a reasonable explanation such as, "The output is 10 more than the input. To find an input from an output, subtract 10. $512 - 10$ is 502.

To find an output from an input, add 10. 890 + 10 = 900.

13 Answer: No; Explanation will vary

Explanation: Answers must include a reasonable explanation such as, "Violet is incorrect because instead of rounding 87 to 90, she should have rounded down to 80. She subtracted the ones (7-4) separately, which left subtraction of the tens (80-30). No regrouping is required for this expression, so the correct decomposition strategy would be (80-30)+(7-4). The correct answer is 53."

14 Answer: D

Explanation: The line is 4 cm long.

15 Answer: B

Explanation: The dog's is 3 paper clips or 6 buttons long.

16 Answer: A

Explanation: 26 centimeters is a good estimate.

17 Answer: 72

Explanation: Answers must include a reasonable explanation such as," The bed is 72 inches because the length of the bed is approximately the length of 4 pillows."

18 Answer: 54

Explanation: Vince scores 26 points, Lori scores 41 points (26 + 15), and Jack scores 54 points (41 + 13).

19 Answer: B

Explanation: Madison's cone is 9 inches tall, Jose's Ice cream cone is 7 inches tall, and Lori's cone is 10 inches tall.

20 Answer: 692

Explanation: To represent the original expression, 6 hundred blocks, 9 tens blocks, and 2 ones blocks should be drawn.

21 Answer: 33

Explanation: Number line should start at 53 and move backwards (left) 20 units to 33.

22   Answer: 100 − 37 = ?; Explanation will vary

Explanation: Answers must include a reasonable explanation such as, "1 hundreds block minus 3 tens blocks and 7 ones blocks."

23   Answer: 400 + 600 + 400 + 400 = 1,800

Explanation: The distance between each large hash mark is 400 miles. Flying to City 1 is 400 miles, City 1 to City 2 is 600 miles, City 2 to City 3 is 400 miles, City 3 to City 4 is 400 miles. The total miles is 1,800.

24   Answer: 5:10

Explanation: "A quarter to five" is 4:45; adding 25 minutes to this time gives 5:10 pm. The hour hand should be between the 5 and the 6. The minute hand should be on 2.

25   Answer: 15

Explanation: The bar graph has a scale of 2, and the bar falls halfway between 14 and 16.

26   Answer: 300

Explanation: 1000 − 500 − 200 = 300 grams

27   Answer: A

Explanation: A cube and rectangular prism have the same number of edges, 12. Four on the top, 4 on the bottom, and 4 on the sides.

28   Answer: 170

Explanation: Wes scores 38 points, Edray scores 51 points (38 + 13), Yani scores 31 points (51 − 20), and Han scores 50 points (31 + 19).

29   Answer: 3

Explanation: Each column is a vertical arrangement of 3 cars.

30   Answer: C

Explanation: There are 16 fish in total. 16 can be represented as the double fact 8 plus 8.

31   Answer: 700

Explanation: 700 is 7 hundreds, 0 tens, and 0 ones.

32   Answer: A

Explanation: 738 is one number directly before 739.

33   Answer: 200 + 70 + 2

Explanation: The picture shows 272 blocks.

34   Answer: D

Explanation: Five quarters is equal to $1.25

35   Answer: 3

Explanation: There are 8 students shorter than 51 inches and 5 students taller than 51 inches. 8 minus 5 is 3.

36   Answer: >

Explanation: 300 is greater than 295. The symbol always points to the smaller side.

37   Answer: B

Explanation: 145 + 87 = 232. Each hash tag is 100. Point B is between 200 and 300. Therefore, B is at 232.

38   Answer: A

Explanation: 130 + 5 = 135 cm.

39   Answer: 23
         Explanations will vary.

Explanation: There are 23 students in Mr. Burr's class. There are 7 students on the playground and 16 (9+7) students in the lunchroom. The segment should start at 0, go to 7, then go 16 units, and end at 23.

40   Answer: No; Explanation will vary

Explanation: Answers must include a reasonable explanation such as," Tommy is wrong because when the minute hand moves from the 6 to the 7, five minutes have passed. Therefore, he should have added 5 not one. The time is 9:35 pm.

prepaze

41   Answer: 2 one-dollar bills, 1 quarter, 1 dime, 2 nickels, 1 penny; Explanation may vary

Explanation: Answers must include a reasonable explanation such as,"Liz has a total of $3.96 in her backpack. I can see $1.50 of it so $3.96 − $1.50 equals $2.46. $2.46 has to be made up with 2 dollar bills (3 -1 = 2). That means there is $0.46 made of 5 coins (9 − 4 = 5). Those 5 coins have to be 1 quarter, 1 dime, 2 nickels, 1 penny."

42   Answer: 135

Explanation: Second grade has 125 students, first grade has 135 (125 + 10 = 135), fourth grade has 145 (135 + 10 = 145), and third grade has 135 (145 − 10 = 135).

43   Answer: Yes, Explanation will vary

Explanation: Answers must include a reasonable explanation such as, "Zaria is correct because 300 − 24 = 276. She regrouped correctly 2 hundreds, 9 tens, and 10 ones is equal to 300 and then subtracted each digit."

44   Answer: Wednesday

Explanation: Wednesday has the most icons in the pictograph. On Wednesday the owl flew 32 miles.

45   Answer: The values shown on this bar graph are 4, 6, 3, 5, 2. When creating the line plot, there should be 5 data points representing the 5 values on the bar graph.

Explanation: The line plot and bar graph show corresponding values on a given scale. The frequency of each data point is denoted with an X on a line plot, where on the bar graph, it is a bar with a length corresponding to the number of data points.

Made in the USA
Monee, IL
27 May 2022

97145638R00118